# PHARMACEUTICAL
# ACHIEVERS

# PHARMACEUTICAL
# ACHIEVERS

## The Human Face of Pharmaceutical Research

by Mary Ellen Bowden
Amy Beth Crow
and Tracy Sullivan

 Chemical Heritage Press
Philadelphia

Printed in the United States of America.

Book text design by Sylvia Barkan and Patricia Wieland
Cover design by Mark Willie, Willie•Fetchko Graphic Design
Composition by Patricia Wieland
Printed by Thomson-Shore, Inc.

For information about CHF publications write
Chemical Heritage Foundation
315 Chestnut Street
Philadelphia, PA 19106-2702, USA
Fax: (215) 925-1954
Website: http://www.chemheritage.org

**Library of Congress Cataloging-in-Publication Data**
Bowden, Mary Ellen.
    Pharmaceutical achievers / by Mary Ellen Bowden, Amy Beth Crow, and
Tracy Sullivan.
        p. cm.
    Includes bibliographical references and index.
    ISBN 0-941901-30-0 (pbk. : alk. paper)
        1. Pharmacologists—Biography. 2. Drugs—Research—History. I. Crow,
Amy Beth, 1976– . II. Sullivan, Tracy. III. Title.
    RM301.25.B69 2003
    615′.1′0922—dc21
    [B]                                                        2001055313

Cover photos, clockwise from top: Gladys Hobby and Alexander Fleming, courtesy Pfizer, Inc.; Fu-Kuen Lin, courtesy Amgen, Inc.; Leo Sternbach, courtesy Hoffmann–La Roche; Percy Julian, courtesy DePauw University Archives and Special Collections; Louis Pasteur, courtesy Edgar Fahs Smith Collection, University of Pennsylvania Library.

# Contents

viii   Acknowledgments

ix   Introduction

## Part I.
## Beginnings through the 1930s

**1.** EARLY PHARMACEUTICALS

2   Introduction to Chapter 1
3   Joseph-Bienaimé Caventou and Pierre-Joseph Pelletier
6   Louis Pasteur
10   Felix Hoffmann
13   Robert L. McNeil, Jr.
17   Paul Ehrlich

**2.** EARLY PHARMACEUTICAL COMPANIES AND THE PURE FOOD AND DRUGS ACT

20   Introduction to Chapter 2
21   George K. Smith
23   Edward Robinson Squibb
26   Harvey Washington Wiley

**3.** VITAMINS

29   Introduction to Chapter 3
30   Albert Szent-Györgyi
33   Robert R. Williams
35   Gladys Ludwina Anderson Emerson
38   Carl Peter Hendrik Dam, Edward Adelbert Doisy, and Karl Paul Gerhardt Link

**4.** ENZYMES, HORMONES, AND NEUROTRANSMITTERS

42   Introduction to Chapter 4
43   Leonor Michaelis and Maud Leonora Menten
47   John Jacob Abel and Jokichi Takamine
52   Frederick Grant Banting, Charles Herbert Best, James Bertram Collip, and John James Rickard Macleod
57   George Henry Alexander Clowes
59   Sir Henry Hallett Dale and Otto Loewi

**5.** SULFA DRUGS

64   Introduction to Chapter 5
65   Gerhard Domagk
68   Daniel Bovet

## Part II.
## World War II to 1960

**6.** ANTIBIOTICS

72 Introduction to Chapter 6
73 Alexander Fleming
76 Howard Walter Florey and Ernst Boris Chain
82 Dorothy Crowfoot Hodgkin
84 Gladys L. Hobby
86 Andrew J. Moyer
88 John Elmer McKeen
90 Selman Abraham Waksman
93 Elizabeth Lee Hazen and Rachel Fuller Brown

**7.** STEROIDS

96 Introduction to Chapter 7
97 Edward Calvin Kendall, Philip Showalter Hench, Tadeus Reichstein, Max Tishler, and Lewis Hasting Sarett
106 Percy Julian, Russell Marker, and Carl Djerassi

**8.** DRUGS AFFECTING THE CENTRAL NERVOUS SYSTEM: ANTIPSYCHOTICS AND TRANQUILIZERS

113 Introduction to Chapter 8
114 Robert Burns Woodward
118 Paul Charpentier, Henri-Marie Laborit, Simone Courvoisier, Jean Delay, and Pierre Deniker
122 Leo Sternbach and Lowell Randall

**9.** NEWER VACCINES

125 Introduction to Chapter 9
126 Jonas Salk and Albert Bruce Sabin
131 Maurice Hilleman

**10.** CLINICAL TESTING AND REFORM OF THE 1906 PURE FOOD AND DRUGS ACT

134 Introduction to Chapter 10
136 Walter Campbell
139 Sir Austin Bradford Hill
142 Frances Oldham Kelsey

## Part III.
## The 1960s to the 1980s

**11.** RECENT CARDIOVASCULAR AND ANTIULCER DRUGS

146 Introduction to Chapter 11
148 James Whyte Black
151 Miguel Ondetti
154 John Duncia and David Carini
157 Alfred W. Alberts, Georg Albers-Schönberg, and Arthur A. Patchett

**12.** RECENT ANTIHISTAMINES
   AND ANTIDEPRESSANTS

163 Introduction to Chapter 12
164 Albert A. Carr
167 Ray W. Fuller, David T. Wong, and Bryan B. Molloy

**13.** CANCER AND AIDS DRUGS

169 Introduction to Chapter 13
170 George Hitchings and Gertrude Elion
173 Irving Sigal, Emilio Emini, Joel Huff, Joseph Vacca, Bruce Dorsey, and Jon Condra

**14.** BIOTECHNOLOGY

178 Introduction to Chapter 14
179 Paul Berg, Herbert W. Boyer, and Stanley N. Cohen
184 César Milstein and Georges Köhler
187 George Rathmann and Fu-Kuen Lin
191 Charles Weissmann
194 William J. Rutter

**Part IV.**
**Finding Tomorrow's Medicines**

197 Introduction to Part IV

**15.** NEW OR REVISITED AREAS FOR
   PHARMACEUTICAL RESEARCH

198 Gene Therapy
198 Antibiotics
199 Orphan Drugs
200 Improved Chemical Entities
200 Drug Delivery

**16.** NEW METHODS OF DRUG DISCOVERY

203 Combinatorial Chemistry
205 Computational Chemistry
205 New Methods of Screening
206 Medicines in the Days after Tomorrow?

207 Bibliography
214 Index

# Acknowledgments

*Pharmaceutical Achievers: The Human Face of Pharmaceutical Research* and the Web-based instructional modules now appearing among the teacher resources on the Chemical Heritage Foundation's Web page (www.chemheritage.org/EducationalServices/tr.html) have been funded in part by the Dibner Fund, the Bristol-Myers Squibb Foundation, the Merck Company Foundation, and the Pfizer Foundation.

The three authors of this publication wish to thank the members of the steering committee for the book—Thomas Althuis (retired, Pfizer Inc.), Bryce Douglas (retired, Smith-Kline Beecham Corporation), and Jeffrey Sturchio (Merck & Company)—who read our early efforts and commented on the general scope and direction of the work. In assembling the story of pharmaceutical innovation in terms of the contributions of individuals, we had at several points—especially as we researched more recent history—to contact some of the scientists who are among the protagonists in this story. They proved very patient with our attempts to translate highly technical discussions into more accessible language. Of course, any scientific errors that appear in the text are our responsibility, not theirs. Our work benefits further from the advice of scientist-historians at the Chemical Heritage Foundation who read parts of the manuscript, especially Leo Slater and John Ceccatti. Leo also drew most of the molecular models that appear in the text, and Mark Michalovic completed that task.

To help readers see the human face behind pharmaceutical developments, generations of workers searched the world for appropriate photographs to illustrate the text. Amy Crow, one of the authors, was the start-up person in this endeavor, followed by Rhonda Goodman, Megan Susnis, and Martha Witte. Laura Wukovitz and Rasheedah Young lent bibliographical skills to the job of rationalizing working bibliographies into a consolidated reference bibliography.

Frances Coulborn Kohler, Patricia Wieland, and Shelley Wilks Geehr of CHF's publications staff maintained an unflagging desire to make *Pharmaceutical Achievers* a literate and visually appealing work. And to keep us working with a will, we had all those high school teachers and college professors inquiring month after month whether the sequel to *Chemical Achievers: The Human Face of Chemical Achievement* (1997) was ready yet. We hope that we have produced a worthy successor to help their students understand the fascination and the complexities of scientific research.

Mary Ellen Bowden
Senior Research Historian

# Introduction

From vaccines and vitamins, to antibiotics and cardiovascular medications of all descriptions, to medicines to treat AIDS, to the latest biotechnological wonders, pharmaceuticals help to make our lives healthier and longer. These contributions were made not by science or industry in the abstract. Rather, they were made by individual human beings working long and hard, supported and encouraged by their families, universities, companies, or governments. It is hoped that accounts of the struggles and triumphs of these individuals will inspire young people to consider careers in the many areas that support pharmaceutical research and development.

Early in our story pharmaceutical achievements were often made by only one or two pharmacists or physicians who were significantly educated in the chemical sciences—an education long characteristic of those fields—or by a few chemists working in conjunction with other professionals more versed in biology. Throughout the last two centuries productive exchange of information between chemical and biological scientists has aided in the creation of new medicines and furthered our understanding of living systems.

Over time the professions associated with pharmaceutical research and development have each become more diverse and at the same time more interrelated, requiring of their practitioners a good understanding of both chemical and biological concepts. The disciplines associated with these professions now include organic, analytical, and medicinal chemistry, biochemistry, and molecular biology as well as physiology, pharmacology, toxicology, bacteriology, virology, enzymology, and clinical medicine, among others. By focusing in most cases on the early stages of research and development, we have slighted but not neglected the role of the chemical engineers and others who scaled up processes originally developed at the laboratory bench so that new medicines would be available in sufficient quantities for clinical testing and subsequent marketing to the public. Our text also carries hints of the important and interesting business and legal careers that have been and are now available in the pharmaceutical industry and the similar careers to be found in government and public-interest groups.

*Pharmaceutical Achievers* focuses on a select group of individuals who helped to develop some of the key pharmaceutical products during the last two hundred years. The individuals selected are representative of the pharmaceutical innovators who have brought about significant changes in health care in modern times. Dozens of other equally significant candidates beckoned for inclusion.

The lives retold here are interesting not only for the knowledge and methods that the scientists used to make their innovations but also for the social contexts that supported the scientists and their ideas. The fundamentally chronological arrangement of this book demonstrates well the progress that has been made over time in our knowledge of the functioning of the healthy body and of disease processes and in the methods scientists have used to find therapies. The chronological approach also displays in historical order the major directions of pharmaceutical research and just when effective therapies for various diseases were first found. The reader is thereby aided in imagining what his or her fate might have been before, for example, the advent of antibiotics to combat infections and infectious diseases or before the introduction of the polio vaccines.

Along with abundant evidence of progress, there is also evidence of enduring themes in the search for new and more effective medicines. In the accounts of the lives and contributions of several nineteenth-century scientists (Chapter 1), some of these themes are first stated. In Western medicine, at least since the days of the ancient Greeks, there has been a strong belief that the body can maintain itself in health and that an effective cure is one that stimulates the body to restore itself. The vaccines that Louis Pasteur, Paul Ehrlich, and others developed were seen to act in just such a manner.

It was also vigorously maintained, with no real contradiction, that curative agents could be found in substances external to the human body. These substances have often been sought in nature, an eternally important storehouse of molecular variety. From the nineteenth century onward pharmaceutical researchers also had available for their use a trove of synthetic compounds, at first coming from the newly established synthetic dye industry. One task that chemical scientists perform in virtually all drug-discovery stories is the isolation and purification of substances, from whatever sources, that have demonstrable effects on the human body.

From early on in the history of pharmaceuticals, chemists suspected that a relationship existed between the biological activity of a substance and the structure of that substance at its most minute level—since the nineteenth century, understood as molecular structure. When a substance that produced some beneficial effects was weak or had toxic or otherwise undesirable side effects, chemists used chemical techniques to modify the substance and its activities.

In the early twentieth century several diseases that had previously proved puzzling in origin were discovered to be caused by vitamin deficiency. The first step generally was the recognition that particular foods appeared to prevent or cure certain diseases. Then scientists had the task of isolating exactly what chemical substances in the foods were responsible for these good results (Chapter 3). Shortly thereafter, determining the molecular structure of vitamins became a high priority in order to understand their functions better and to develop syntheses if extraction from natural sources proved too expensive.

At about the same time, scientists were busy investigating the roles of others of the body's own chemicals in regulating its functions and transmitting information among its parts (Chapter 4). In several instances drugs from traditional Western medicine were used as probes to understand how living systems work on the chemical level. Although expectations were high that the knowledge obtained would prove invaluable in employing the body's own substances to correct various disease conditions, in only a few instances were these hopes fulfilled in the period before World War II—most dramatically in the administration of epinephrine (adrenaline) and insulin.

The sulfa drugs (Chapter 5), which appeared on the market in 1935, provide prime examples of external agents originally derived from synthetic dyes that, when introduced into the body, proved effective in curing diseases—in this case for infections and several bacterial diseases for which no vaccines had been developed. In many ways the sulfa drugs prepared the way for the introduction during World War II of antibiotics, which like the sulfa drugs are agents external to the human body and produced by microbes for their own purposes (Chapter 6).

The themes already evident in the development of pharmaceuticals before World War II were reinforced in the 1950s. Largely through advances in chemical synthesis, hormone therapies (Chapter 7), including contraceptives, became available to substantial numbers of people at affordable prices. In this same period, on the basis of accumulating knowledge of neurotransmitters, scientists developed the first really effective drugs to treat mental illness—without reproducing the worst effects of earlier drugs (Chapter 8). The mumps-measles-rubella and the polio vaccines also date from

the 1950s. This era differed from earlier periods of vaccine development because of the expanded possibilities for culturing disease-carrying microbes in the laboratory and the more elaborate clinical trials in which vaccines were tested (Chapter 9).

For investigators early in our story determining the safety and efficacy of a new drug was a rather informal and often risky business. Investigators who sometimes tried new preparations on themselves were placed in danger. The general public was often put at risk by manufacturers who sold them drugs despite knowing that some were dangerous or that their effects were unknown. By the turn of the nineteenth century those pharmaceutical manufacturers who had built up good reputations for high-quality products agreed with reform-minded physicians and government officials that drugs for which extravagant and erroneous claims were made should be removed from the market. The result was the Pure Food and Drugs Act of 1906 (Chapter 2). Meanwhile, more rigorous means of testing new drugs, for their safety as well as their efficacy, were being developed in the pharmaceutical community. These methods of testing were eventually incorporated into a new act and amendments to it (Chapter 10).

Along with the discovery and testing of new drugs came a greater understanding of the body's own functions, down to the molecular level. In the 1960s it became possible to target cells and molecules within particular physiological processes whose function might be blocked by appropriate compounds. Searches were done in libraries of compounds, both natural and synthetic, that pharmaceutical companies and other institutions had collected. Once a compound was found that exhibited the kind of activity desired—however weakly—it was selected as the "lead" candidate. Then its structure was modified much more systematically than was possible in earlier times in order to convert it into a safe, effective, and easy-to-administer medicine. Closer to the present, computerized databases and molecular-modeling software programs have become integral to such searches. Since the 1960s many successful medications have been developed following the principles of this method, so-called rational design, including commonly prescribed cardiovascular and antiulcer drugs (Chapter 11), antihistamines and antidepressants (Chapter 12), and cancer and AIDS drugs (Chapter 13).

These drugs were each intended to interact with specific

receptors or enzymes or even the cell's own method of re-producing itself and its synthesis of life-sustaining sub-stances. From parallel and related investigations of genetics and enzymes, combined with "rational drug design," arose the application of biotechnology to pharmaceuticals (Chapter 14). Through recombinant DNA processes and hybridoma technology, scientists have already succeeded in duplicating a number of the body's own chemicals, which can in many cases be used as self-curative agents.

Beginning during World War II but growing ever larger thereafter, the average number of researchers involved in a given project has tended to increase, although the passage of time or another mechanism of selection, such as choosing winners of a prize, has commonly whittled down the number of scientists whose names are remembered in connection with a particular discovery. As with any creative effort it is common to find within these groups instances of personal and professional rivalry. At least as common are instances of rivalries between groups, often disputes over patents. Because the knowledge base from which scientists at various companies, universities, or government agencies draw is nearly the same for all, simultaneous or nearly simultaneous discoveries are highly likely. Moreover, the economic stakes are so high that litigation is bound to result. But these disputes make the stories of the lives of the achievers even more interesting as records of *human* endeavor—which pharmaceuticals research is in essence.

In a concluding section (Chapters 15 and 16) we outline the areas for pharmaceutical research that have most recently opened or reopened as well as some of the newer methods of conducting this research. Here the near-sightedness entailed in proximity to the present makes naming achievers difficult, so we have chosen not to do so. And not doing so may have the added benefit of making it easier for young people learning of these developments to imagine themselves as pharmaceutical achievers of the future.

While much has changed in recent times in the conduct of pharmaceutical research in terms of its knowledge base, scale, and organization, much has remained the same. The human qualities required for participating in the grand quest to prevent the spread of disease, cure sickness, and alleviate pain and suffering are perhaps the most constant of all: intelligence, curiosity, energy, and perseverance.

# Part I.

# Beginnings
## through
## the
## 1930s

Clockwise from top left: Frederick Banting and Charles Best (courtesy Thomas Fisher Rare Book Library, University of Toronto); Gladys Emerson (courtesy Merck Archives, Merck & Co., Inc.); Louis Pasteur (courtesy Edgar Fahs Smith Collection, University of Pennsylvania Library); Jokichi Takamine (courtesy Bayer Corporation); Gerhard Domagk (courtesy Bayer Corporation).

# 1. Early Pharmaceuticals

From time immemorial humans have sought relief from pain
and disease in various plant, animal, and mineral substances.
But by the nineteenth century, scientists were already using
strategies for finding new treatments and improving their
safety and effectiveness that are clear precursors of those used
in today's pharmaceuticals research and development.

One such strategy was to use chemical methods to isolate the pure active ingredients from natural materials—mainly plant materials—known to have beneficial effects. Prompting this strategy were two drawbacks of remedies based on natural sources. First, the amount of active ingredients—those of interest to pharmacists and physicians—in the natural materials varies unpredictably, making it impossible to prepare and prescribe doses with predictable results. Second, because such materials contain many different chemicals, they often produce unwanted side effects.

One early success of this chemical strategy was the isolation of quinine from the bark of the cinchona tree, the most successful treatment for malaria until the advent of antibiotics. Another was the isolation of salicylic acid from the bark of the willow tree, a traditional pain reliever (analgesic) and fever reducer (antipyretic). The sequel to this isolation illustrates another chemical strategy: that of modifying naturally active compounds to improve their qualities as drugs. Thus salicylic acid proper was transformed by chemistry to acetylsalicylic acid, or aspirin.

Beginning in the nineteenth century, chemists also had available a host of new organic chemicals for use and modification, mainly coming from the new synthetic-dye industry and based on coal-tar derivatives. Another popular pain reliever, Tylenol, was born in such a chemical context.

Its delayed development shows the important role in the pharmaceutical discovery process of scientists who test the effects of medicines on animal tissue, on animals, and on humans.

Two of the most revered medical scientists of all time, Louis Pasteur and Paul Ehrlich, appear in this chapter. In Pasteur we see an early proponent of the germ theory of disease, which informs much of pharmaceutical history. Ehrlich and Pasteur together represent another major theme of pharmaceutical history, the attempt to rally the body's own natural defenses against disease—in Pasteur's case by developing vaccines from live but attenuated vaccines. Ehrlich, too, developed vaccines a generation later, but from the sera of immunized animals. In their work they also displayed facets of the emerging understanding of the body's immune system—an understanding that in its intimate details is still being improved today.

In his work Ehrlich succeeded in combining the two themes of using the body's own resources and intervening with remedies drawn from sources external to the body to combat disease. In his efforts to understand vaccines in chemical terms, he theorized that chemical structures exist in the body's own substances like those he knew were present in synthetic dyes. This theory in turn enabled him to conduct a prolonged search for the first effective treatment for syphilis—salvarsan.

## Joseph-Bienaimé Caventou (1795–1877) and Pierre-Joseph Pelletier (1788–1842)

In the early nineteenth century, French pharmacists Joseph-Bienaimé Caventou and Pierre-Joseph Pelletier isolated a number of active principles from plant materials used in drugs—that is, particular chemical compounds producing physiological effects in patients. These drugs, often of exotic origin and relatively recently introduced into European medical practice, were originally administered as brews made from ground flowers, leaves, roots, and bark of trees and other plants. Then—as now—isolating active principles was undertaken to achieve higher potency, fewer side effects, and more consistency than would be available directly from unrefined natural materials.

Caventou and Pelletier shared a common heritage as the sons of pharmacists and graduates of the rigorous Parisian system of educating members of their profession. The required curriculum at the École de Pharmacie included a strong component of chemistry, based on Antoine Lavoisier's reform of the subject. At the time of their discoveries Pelletier was already on the teaching staff of the École and operating a pharmacy in Paris, and Caventou would soon be similarly employed. Pelletier owned a chemical plant in nearby Clichy as well.

Pelletier's first collaborator was François Magendie, a physician now revered as a pioneer experimental physiologist. As a physician Magendie was interested in finding new and improved medicines for his patients. As a physiologist he used the newly discovered substances as probes to investigate how the nervous system and other systems of living organisms function.

Working in Pelletier's laboratory, Pelletier and Magendie conducted chemical investigations of a wide variety of natural materials used in medicines of the day, including the root of the ipecacuanha (or ipecac) plant, the basis for a purge of the same name that was commonly prescribed. The two were joined in their search for ipecac's active principle by Caventou, who was then a twenty-two-year-old pharmacy intern. Using himself as a test subject, Caventou swallowed six grains of a vile-smelling fatty constituent of ipecac; when this noxious stuff did not induce vomiting, it was thus ruled out as the active ingredient in the medicine. Another substance that they isolated worked as expected on Magendie, Pelletier, and some students; they named it "emetine" from the Greek word for *vomiting*.

Emetine is one of the alkaloids, a group of basic organic substances (those that form salts with acids) with complex molecular structures that are among the most physiologically active substances known to mankind. The first alkaloid to be identified was morphine, extracted by Friedrich Wilhelm Sertürner in 1805 (research published in 1817) from opium derived from the opium poppy; up until that time all vegetable matter was thought to be acidic.

In an amazing five years, from 1817 to 1821, Caventou and Pelletier not only studied and named the green pigment in leaves "chlorophyll," but they also isolated and characterized a number of alkaloids. Among these were the poison strychnine, a derivative of St. Ignatius's beans (*Strychnos ignatii*) and the seeds of nux vomica (*S. nux vomica*), known to the English as "Quaker buttons"; quinine from the bark of the cinchona tree—an effective treatment for malaria, for which they received a prize from the Académie des Sciences (see also Robert Burns Woodward, p. 114); and caffeine—the popular stimulant present in coffee, tea, and today's cola drinks.

With the purest samples of these and other new substances that they had obtained through chemical manipulations, Caventou and Pelletier attempted to use combustion analysis to determine empirical chemical formulas. In all of the substances they found varying proportions of carbon, hydrogen, and oxygen. But it was not until 1823, when Pelletier and Jean-Baptiste Dumas repeated such analyses for nine alkaloids, that it was recognized that the alkaloids also contain nitrogen. Determining the molecular structure of and synthesizing most alkaloids awaited the twentieth century.

■ Joseph-Bienaimé Caventou. Oil painting. After conservation 1998. Collection of the Wellcome Library, London. ©The Trustee of the Wellcome Trust, London.

ROSSELIN, éditeur, 21, quai Voltaire.

Lith. de Grégoire et Deneux, 9, rue Cassette, à Paris.

■ Pierre-Joseph Pelletier. Courtesy Edgar Fahs Smith Collection, University of Pennsylvania Library.

# Louis Pasteur (1822–1895)

Louis Pasteur is revered by his successors in the life sciences as well as by the general public. In fact, his name provided the basis for a household word—*pasteurized.* He demonstrated that microorganisms cause both fermentation and disease, which weighed heavily on the side of germ theory at a time when its validity was still being questioned. He developed the earliest vaccines against fowl cholera (unrelated to human cholera); anthrax, a major livestock disease (which in recent times has been used against humans in germ warfare); and the dreaded rabies.

Pasteur was the middle child of five in a family that had for generations been tanners. Young Pasteur's gifts seemed to be more artistic than academic until near the end of his years in secondary school. Spurred by his mentors' encouragement, he undertook rigorous studies to compensate for his academic shortcomings in order to prepare for the École Normale Supérieure, the famous teachers' college in Paris. He earned his master's degree there in 1845 and his doctorate in 1847.

While waiting for an appropriate appointment, Pasteur continued to work as a *préparateur* (laboratory assistant) at the École Normale. There he made further progress on the research he had begun for his doctoral dissertation—investigating the ability of certain crystals or solutions to rotate plane-polarized light clockwise or counterclockwise, that is, to exhibit "optical activity." He was able to show that in many cases this activity related to the shape of the crystals of a compound. He also reasoned that there was some special internal arrangement to the molecules of such a compound that twisted the light—an "asymmetric" arrangement.

Pasteur secured his academic credentials with scientific papers on this and related research and was then appointed in 1848 to the faculty of sciences in Strasbourg and in 1854 to the faculty in Lille. There, Pasteur launched his studies on fermentation—an interest that traced to his search while at Strasbourg for a good supply of amyl alcohol, an optically active substance produced in the fermentation of sugar beets. In attempting to determine the source of such optical activity, he discarded the usual answer that the sugar that served as the starting material for the fermentation was also optically active. According to Pasteur, there is a necessary connection between asymmetry and life, and so he sided with

the minority view among his contemporaries that each type of fermentation is carried out by a living microorganism. At the time the majority believed that fermentation is simply a series of chemical reactions in which enzymes—themselves not yet securely identified with life—play a critical role.

In 1857 Pasteur returned to the École Normale as director of scientific studies. In the modest laboratory that he was permitted to establish there, he continued his study of fermentation and fought long, hard battles against the theory of "spontaneous generation." Figuring prominently in early rounds of these debates were various applications of his pasteurization process, which he originally invented and patented (1865) to fight the "diseases" of wine. He realized that these were caused by unwanted microorganisms that could be destroyed by heating wine to a temperature between 60 and 100° C. The process was later extended to all sorts of other spoilable substances.

At the same time Pasteur began his fermentation studies, he adopted a related view on the cause of diseases. He and a minority of other scientists believed that diseases arose from the activities of microorganisms. Opponents believed that diseases, particularly major killer diseases, arose in the first instance from a weakness or imbalance in the internal state and quality of the afflicted individual. In an early foray into the causes of particular diseases, in the 1860s, Pasteur was able to determine the cause of the devastating blight that had befallen the silkworms that were the basis for France's then-important silk industry. Surprisingly, he found that the guilty parties were two microorganisms rather than one.

Pasteur did not, however, fully engage in studies of disease until the late 1870s, after several cataclysmic changes had rocked his life and that of the French nation. In 1868, in the middle of his silkworm studies, he suffered a stroke that partially paralyzed his left side. (Afterward he had to rely more on his loyal assistants, of whom he had many over the years, to carry out the experiments he devised.) Soon thereafter, in 1870, France suffered a humiliating defeat at the hands of the Prussians, and Emperor Louis Napoleon was overthrown. Nevertheless, Pasteur successfully concluded with the government of the Third Republic the

■ Louis Pasteur, at the age of sixty-three, as painted by Finnish artist Albert Edelfelt. Courtesy Edgar Fahs Smith Collection, University of Pennsylvania Library.

■ Glassware of the same type Louis Pasteur would have used to culture microorganisms. From the Pasteur Memorial Collection, Fisher Collection, Chemical Heritage Foundation. Photo by Gregory Tobias.

negotiations he had begun with Louis Napoleon. The government agreed to build a new laboratory for him, to relieve him of administrative and teaching duties, and to grant him a pension and a special recompense in order to free his energies for studies of diseases.

In his research campaign against disease Pasteur first worked on expanding what was known about anthrax, but his attention was quickly drawn to fowl cholera—an investigation that led to his discovery of how to make vaccines by attenuating, or weakening, the microbe involved. At the end of the eighteenth century Edward Jenner had introduced inoculation with cowpox as protection against human smallpox—the first vaccine (from the Latin, *vaccus,* cow). But since then no one had been able to provide an artificial means of protecting potential victims from contracting diseases. Pasteur usually refreshed the laboratory cultures he was studying—in this case, fowl cholera—every few days; that is, he returned them to virulence by reintro-

ducing them into laboratory chickens with the resulting onslaught of disease and the birds' death. Months into the experiments, Pasteur let cultures of fowl cholera stand idle while he went on vacation. When he returned and the same procedure was attempted, the chickens did not become diseased as before. Pasteur could easily have deduced that the culture was dead and could not be revived, but instead he was inspired to inoculate the experimental chickens with a virulent culture. Amazingly, the chickens survived and did not become diseased; they were protected by a microbe attenuated over time.

Realizing he had discovered a technique that could be extended to other diseases, Pasteur returned to his study of anthrax. Here he had to modify his attenuation process to accommodate a microbe that has a spore stage. Pasteur produced vaccines from weakened anthrax bacilli that could indeed protect sheep and other animals. In public demonstrations at Pouilly-le-Fort before crowds of observers, twenty-four sheep, one goat, and six cows were subjected to a two-part course of inoculations with the new vaccine, on 5 May 1881 and again on 17 May. Meanwhile a control group of twenty-four sheep, one goat, and four cows remained unvaccinated. On 31 May all the animals were inoculated with virulent anthrax bacilli, and two days later, on 2 June, the crowd reassembled. Pasteur and his collaborators arrived to great applause. The effects of the vaccine were undeniable: the vaccinated animals were all alive. Of the control animals all the sheep were dead except three wobbly individuals who died by the end of the day, and the four unprotected cows were swollen and feverish. The single goat had expired too.

Pasteur then wanted to move into the more difficult area of human disease, in which ethical concerns weighed more heavily. He looked for a disease that afflicts both animals and humans so that most of his experiments could be done on animals (although here too he had strong reservations). Rabies, the disease he chose, had long terrified the populace, even though it was in fact quite rare in humans. Up to the time of Pasteur's vaccine a common treatment for a bite by a rabid animal had been cauterization with a red-hot iron in hopes of destroying the unknown cause of the disease, which almost always developed anyway after a typically long incubation period.

Rabies presented new obstacles to the development of a successful vaccine, primarily because the microorganism causing the disease could not be specifically identified nor could it be cultured in vitro (Latin for "in glass," meaning in the laboratory and not in an animal). As with other infectious diseases, rabies could be injected into other species

and attenuated. Attenuation of rabies was first achieved in monkeys and later in rabbits. Meeting with success in protecting dogs, even those already bitten by a rabid animal, on 6 July 1885, Pasteur agreed with some reluctance to treat his first human patient, Joseph Meister, a nine-year-old who was otherwise doomed to a near-certain death. Success in this case and thousands of others convinced a grateful public throughout the world to make contributions to the Institut Pasteur. It was officially opened in 1888 and continues to this day as one of the premier institutions of biomedical research in the world. Its tradition of discovering and producing vaccines is carried on today by the pharmaceutical company Aventis Pasteur.

Pasteur's career shows him to have been a great experimenter, far less concerned with the theory of disease and immune response than with dealing directly with diseases by creating new vaccines. Still it is possible to discern his notions on the more abstract topics. Early on he linked the immune response to the biological, especially nutritional, requirements of the microorganisms involved; that is, the microbe or the attenuated microbe in the vaccine depleted the food source during its first invasion, making the next onslaught difficult for the microbe. Later he speculated that microbes could produce chemical substances toxic to themselves that circulated throughout the body, thus pointing to the use of toxins and antitoxins in vaccines. He lent support to another view by welcoming to the Institut Pasteur Élie Metchnikoff and his theory that "phagocytes" in the blood—white corpuscles—clear the body of foreign matter and are the prime agents of immunity.

# FELIX HOFFMANN (1868–1946)

Within a two-week period in August 1897 the German chemist Felix Hoffmann synthesized two drugs: aspirin, one of the most widely beneficial drugs ever, and heroin, one of the most harmful of illegal substances.

These two drugs represent the efforts of late-nineteenth-century chemists to create new substances that could be used as medicines, not just to isolate active principles from natural products or to imitate them. One approach toward this end was to modify known physiologically active substances; another was to perform chemical operations on one or more of the myriad organic compounds created as products or by-products of the synthetic dye industry that had developed in the nineteenth century and that was particularly strong in Germany.

Felix Hoffmann was the son of a manufacturer in the town of Ludwigsburg in Swabia, Germany. He first found employment in pharmacies in various cities and towns around Germany and later studied chemistry and pharmacy at the University of Munich, graduating in 1893. Recommended by one of his professors, Adolf von Baeyer (who would win the Nobel Prize in chemistry in 1905 for his work in synthesizing dyes), Hoffmann joined the newly established pharmaceutical research department at the Bayer Company in Elbersfeld.

In the summer of 1897 Hoffmann was acetylating (adding the acetyl group, $CH_3CO$, to) all sorts of molecules, with hopes of improving the strength or decreasing the toxicity

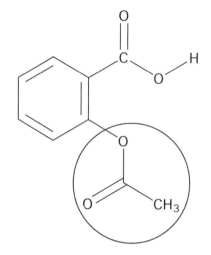

**Figure 2.** Acetylsalicylic acid (aspirin).

of physiologically active substances. This strategy had worked with Bayer's first medicines: the fever-reducing phenacetin (1888), the acetylated form of *p*-nitrophenol, itself a useless by-product of manufacturing blue dyestuffs; and the anti-diarrheal called Tannig (1894), which was acetylated tannic acid, a component of substances long used in tanning leather. As Hoffmann recalled twenty years later, the Bayer chemists worked by instinct and talked about having a "good nose" for discovery.

Legend has it that Hoffmann was searching for a medicine to ease his father's rheumatic pains when he acetylated salicylic acid, the active principle in salves and teas made from willow bark and certain other plant materials. Since antiquity the pain-relieving and fever-reducing properties of willow bark were well known, and in the early nineteenth century salicylic acid was isolated from the bark by several chemists. In 1859 Hermann Kolbe determined its chemical structure and synthesized it (Figure 1). In 1874 the Heyden Company near Dresden began manufacturing and selling synthetic salicylic acid, a cheaper product than the extract from willow bark itself. However, salicylic acid had unpleasant side effects: it irritated the stomach, and some patients were simply unable to tolerate it.

As soon as Hoffmann succeeded in acetylating salicylic acid to produce acetylsalicylic acid (Figure 2), Heinrich

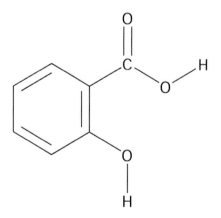

**Figure 1.** Salicylic acid.

■ Felix Hoffmann. Courtesy Bayer Corporation.

Dreser, the head of Bayer's pharmaceutical laboratory, tested the substance for toxicity on himself. Then he set up a series of animal experiments. Tests on people—patients in a hospital in Hall an der Saale—began soon afterward. Acetylsalicylic acid was given the name "Aspirin," from the *A* for *acetyl* and the *spirin* from *Spirea*, the genus name for shrubs that are an alternative source of salicylic acid. An application for a German patent was rejected, because in fact acetylsalicylic acid was not a new substance; it had first been synthesized in 1853 by the French chemist Charles Gerhardt, in impure form, and later by the German chemist Carl J. Kraut, in crystalline form.

Today there is debate as to whether this account is complete. Some evidence has surfaced that indicates that Arthur Eingruen, another Bayer employee, played a significant role in the development of aspirin, and it has been suggested that he was left out of the story as it has been told since the 1930s because he was Jewish.

Regardless, the Bayer Company, recognizing that it had a potential blockbuster in aspirin, aggressively marketed the drug worldwide. In the United States, Bayer was able to obtain a patent, giving the company the monopoly on manufacturing the drug from 1900 to 1917. When Bayer's American plants were sold in 1919 as part of the reparations exacted from Germany after World War I, Sterling Products of Wheeling, West Virginia, was willing to invest the then unheard-of sum of $3 million for Bayer's U.S. drug properties. But Sterling was unable to protect the trademark status of aspirin. It thus became a staple of the over-the-counter market in the United States and elsewhere. More than one hundred years after its invention it continues to be a popular drug, with uses extending far beyond those envisaged by its original creators. SmithKline Beecham eventually purchased Sterling's worldwide over-the-counter pharmaceutical business and in turn sold the U.S. portion, including aspirin, to Bayer for a sum of $1 billion.

Heroin is another story. Dreser, while still a professor in Göttingen, had worked on the effect of codeine, a weaker derivative of opium than morphine, on breathing. He instructed Hoffmann to acetylate morphine with the objective of producing codeine; the result instead was a substance that was named "heroin." But the same compound had already been discovered in 1874 by the English chemist C. R. A. Wright and so was unpatentable. Before the extreme addictiveness of heroin was recognized, however, it was widely sold to suppress heavy coughs, to relieve the pain of childbirth and serious war injuries, to prepare patients for anesthesia, and to control certain mental disorders. Since the 1930s it has been banned in most countries.

■ The interior of McNeil's Drug & Prescription Store at Front and York Streets, Philadelphia, 1900. Courtesy McNeil Consumer Healthcare.

## ROBERT L. MCNEIL, JR. (1915– )

Even before aspirin was created, the active ingredient in one of today's most popular pain relievers and fever reducers, Tylenol, had been discovered. But because toxicology and pharmacology were still in their early stages in the late nineteenth century, this substance—*N*-acetyl *p*-aminophenol—was left on the shelf for over fifty years, until research in the mid-twentieth century demonstrated its efficacy and safety. In the United States its popularity can be traced to a decision made in 1954 by the New Products Committee at McNeil Laboratories to undertake the additional laboratory and clinical testing needed to gain approval for the drug from the U.S. Food and Drug Administration and then to market it.

In 1886, at the University of Strassburg, then located in Germany, two young chemists, Arnold Cahn and Paul Hepp, were asked by their professor to use the coal-tar derivative naphthalene as an "internal antiseptic" to treat a patient suffering from worms and a variety of other ailments. The patient's fever subsided—surprisingly, since naphthalene was not known to have fever-reducing, or antipyretic, properties. Cahn and Hepp suspected that the pharmacist had made an error, as indeed he had: he had sent by mistake another simple organic chemical, also derived from coal tar, acetanilide. They published a report of their finding, and a small company near Frankfurt, Kalle and Company, later taken over by Hoechst, began to manufacture a drug named Antifebrin, whose active ingredient was acetanilide.

When Antifebrin first came on the market, the Bayer Company in Elberfeld, Germany, began to look for a similar drug. Oscar Hinsberg, a chemist at Bayer, synthesized the ethyl ether of acetanilide by acetylating $p$-nitrophenol, and Bayer sold the drug, trademarked as phenacetin, from 1888 on as an antipyretic and analgesic, or pain reliever. Almost immediately physiological chemists investigating the relationships between the chemical composition and the physiological function of Antifebrin and phenacetin discovered through urinalyses that the metabolic processes of animals and humans transformed these drugs into $N$-acetyl $p$-aminophenol.

In 1893 Joseph von Mering, who was then conducting his famous research on diabetes (see Frederick Grant Banting et al., p. 52), published a clinical study of the effects of several compounds of the aminophenol structural class. Among other conclusions, he wrote (wrongly, as would later be determined) that $p$-aminophenol and $N$-acetyl $p$-aminophenol were active antipyretics but were not suitable for clinical use. One of the relatively few patients he tested with $p$-aminophenol turned blue, showing that the capacity of the patient's red blood cells to distribute oxygen to the body had been destroyed, in a condition called methemoglobinemia. In the case of $N$-acetyl $p$-aminophenol he reported no specific observations of patients, information that he had provided for the other substances he had tested.

Meanwhile acetanilide and phenacetin (which lost its status as a trade name) competed with aspirin through World War II. In 1938 the U.S. Food and Drug Administration, newly strengthened by the Food, Drug, and Cosmetic Act of that year, briefly pulled acetanilide from the market because they suspected it caused agranulocytosis, an abnormal condition of the white blood cells.

Although the law did not require companies to test products already on the market, Walter Ames Compton, director of research at Miles Laboratories in Elkhart, Indiana, recognized the critical importance of using the latest methods of pharmacology and toxicology to investigate the standard remedies like Miles's Alka-Seltzer, created before such tools were available. He also realized that in the course of time these new investigations might well lead to new or improved drugs. In 1939 he convinced Miles Laboratories and nine other companies to establish an independent institute, the Institute for the Study of Analgesic and Sedative Drugs, to conduct basic research. Although these companies had their own research laboratories, they agreed that the results generated by an outside body would be more credible to the public.

The newly founded institute hired scientists from leading medical schools to conduct investigations, among them a team investigating acetanilide and phenacetin: Leon Greenberg and David Lester of Yale University and Bernard Brodie, Julius Axelrod, and Frederick B. Flinn of New York University. (During the course of the research Brodie and Axelrod moved to the National Heart Institute at the National Institutes of Health.) Their reports showed conclusively that in humans the active metabolite of both acetanilide and phenacetin was $N$-acetyl $p$-aminophenol and that this compound accounted for the antipyretic and analgesic properties of the two drugs. Methemoglobinemia, it was found, was caused not by $N$-acetyl $p$-aminophenol but by another metabolic breakdown product of the drugs under study. The researchers also cleared all three compounds of the suspicion of causing agranulocytosis. In short, $N$-acetyl $p$-aminophenol appeared to be free from side effects that would preclude its use in normal dosage. These findings and others were presented at a one-day symposium sponsored by the Institute for the Study of Analgesic and Sedative Drugs and held in May 1951 at the Biltmore Hotel in New York City. Earlier, good reports coming out of Yale and New York Universities had led several American pharmaceutical companies to begin manufacturing products containing $N$-acetyl $p$-aminophenol, but their sales efforts were halfhearted, largely because they feared that the new products would adversely affect sales of their aspirin-based products.

The drug truly emerged from the shadows soon thereafter, when, at an American Pharmaceutical Manufacturers Association meeting, Robert L. McNeil, Jr., then vice president of McNeil Laboratories, learned more of the institute's research results in an informal discussion with Raymond L. Conklin, who was vice president of the institute as well as vice president of Ames Company, a subsidiary of Miles Laboratories, in Elkhart, Indiana. That discussion put McNeil Laboratories on a course that resulted in its all-time best-seller, Tylenol.

■ Robert McNeil, Jr., in 1956. Courtesy Robert L. McNeil, Jr. Photo by Tom McCaffrey.

Robert L. McNeil, Jr., represented the third generation of his family at McNeil Laboratories. The company had evolved from a neighborhood pharmacy established in Philadelphia in 1879. Until the 1930s it had specialized in selling hundreds of drugs to physicians who dispensed them without prescription to their patients. After earning a B.S. in physiological chemistry and bacteriology at Yale University in 1936, Robert McNeil entered the business in 1937, while completing a second bachelor's degree at the Philadelphia College of Pharmacy, now University of the Sciences in Philadelphia. (He simultaneously pursued graduate studies at Temple University.) He soon was made responsible for reevaluating the company's product line, reorganizing its departments, and creating a research-and-development division to develop prescription products that would comply with the "safe for use" requirements of the federal Food, Drug and Cosmetic Act of 1938 (see Walter Campbell, p. 136).

When McNeil returned from the pharmaceutical manufacturers meeting, he had Charles F. Kade, Jr., the head of the firm's medical sciences division, and James M. Shaffer, head of its clinical investigation division, confirm the significance of Ray Conklin's verbal report on $N$-acetyl $p$-aminophenol. He then convened a meeting of the firm's New Products Committee, which included these scientists as well as his brother Henry McNeil, then vice president of marketing, and the directors of sales (David S. Lamont), advertising and promotion (John Hogan), and market planning (Douglas G. Lovell, Jr.). It was Lovell who subsequently coined the name *Tylenol* from the drug's chemical name, and Robert McNeil who suggested *acetaminophen* as the generic term.

Presented with a proposal to put $N$-acetyl $p$-aminophenol on the market, the marketing and sales executives on the New Products Committee raised the obvious question: why try to sell a drug to compete with aspirin that would cost more than aspirin? McNeil convinced the doubters by pointing out that the drug had antipyretic and analgesic properties similar to those of aspirin but did not cause the stomach irritation that aspirin often did, especially when taken too frequently—a property that had only been touched on at the symposium in New York. Furthermore, McNeil planned to aim at a different niche: developing a drug that the marketing division could promote to physicians for them to prescribe—not an over-the-counter product that would compete directly with aspirin.

Unlike many other pharmaceutical companies, McNeil Labs did not have its own aspirin product and did not have to fear such competition. For example, Sterling-Winthrop decided to rely solely on aspirin in the United States rather than adopt Panadol, an $N$-acetyl $p$-aminophenol product that its subsidiary, Bayer Ltd., had developed for the British market on the basis of the Yale and Columbia studies.

As the McNeil plan evolved, the pediatric field was targeted, with the objective of developing a liquid-dosage form for children. First McNeil's pharmacology department, headed by David Marsh, confirmed the pharmacologic and toxicologic conclusions of the Lester-Greenberg and Brodie-Axelrod-Flinn groups and others. Clinical trials of a pediatric elixir were conducted in cooperation with pediatricians Donald A. Cornely and Joseph A. Ritter. James Roth and other gastroenterologists studied the local erosion effect of aspirin's hydrolyzing in the stomach to acetic acid and salicylic acid—and demonstrated it in vivo to the marketing executives! The pharmaceutical development department, headed by Albert Mattocks, prepared a special solvent system for the elixir that kept the active ingredient in solution but was low in alcohol and palatable to young patients. The marketing group then proceeded with plans to introduce Elixir Tylenol, using the slogan "for little hotheads" in the caption for the outer carton, which was shaped like a fire engine. Six months after the decision by the New Products Committee to proceed with $N$-acetyl $p$-aminophenol, the U.S. Food and Drug Administration granted its approval, and Elixir Tylenol was introduced in June 1955.

Not long afterward, McNeil Laboratories was purchased by Johnson & Johnson, which had grown since its founding in 1887 into one of the nation's biggest consumer and health-products companies. As part of Johnson & Johnson, McNeil Labs developed and gained FDA approval for a line of Tylenol with codeine, which eventually became the most frequently prescribed of any pharmaceutical brand, followed by other dosage forms and combinations of Tylenol for sale to hospitals and over the counter in drugstores.

## Paul Ehrlich (1854–1915)

In 1906 Paul Ehrlich prophesied the role of modern-day pharmaceutical research, predicting that chemists in their laboratories would soon be able to produce substances that would seek out specific disease-causing agents—"magic bullets," as he called them. Ehrlich himself met with signal successes in the emerging fields of serum antitoxins and chemotherapy.

Ehrlich was born near Breslau—then in Germany, but now known as Wrocław, Poland. He studied to become a medical doctor at the university there and in Strasbourg, Freiburg im Breisgau, and Leipzig. In Breslau he worked in the laboratory of his cousin Carl Weigert, a pathologist who pioneered the use of aniline dyes as biological stains. Ehrlich became interested in the selectivity of dyes for specific organs, tissues, and cells, and he continued his investigations at the Charité Hospital in Berlin. After he showed that dyes react specifically with various components of blood cells and the cells of other tissues, he began to test the dyes for therapeutic properties to determine whether they could kill off pathogenic microbes. Classmates and others recalled Ehrlich as the man with blue, yellow, red, and green fingers.

After a bout with tuberculosis—probably contracted in the laboratory—and his subsequent cure with Robert Koch's tuberculin therapy, Ehrlich focused his attention on bacterial toxins and antitoxins. At first he worked in a small private laboratory, but then he was invited to work at Koch's Institute for Infectious Diseases in Berlin. The post-Pasteur world was an exciting time to be looking for cures and preventives, and Koch's Institute was one of the best places to be. Among Ehrlich's new colleagues were Emil von Behring and Shibasaburo Kitasato, a Japanese scientist whose fellow countrymen were to play a critical role in Ehrlich's research. Von Behring and Kitasato had recently developed sera therapies—first for tetanus, then for diphtheria (for which Behring would win the first Nobel Prize in physiology or medicine)—and evolved the concept of antitoxin to explain the immunizing properties of sera. Whereas Pasteur's vaccines and Koch's tuberculin were made from weakened bacteria, von Behring used the serum, or cell-free blood liquid, from naturally or artificially immunized animals to induce immunity. One of Ehrlich's jobs at the institute was to make von Behring's diphtheria antitoxin in quantity and later to review the quality of the product produced by the chemical-pharmaceutical company Hoechst. In carrying out this work, he learned how to boost immunity systematically and how to produce high-grade sera. Although Ehrlich and von Behring often quarreled, they maintained a lifelong relationship, for better or worse.

In recognition of Ehrlich's accomplishments and of his promise as a researcher, in 1896, the Institute for Serum Research and Serum Testing was established for him in a Berlin suburb. In 1899 the institute, which was originally housed in a one-story ramshackle building, was moved to the city of Frankfurt to more suitable quarters and renamed the Royal Prussian Institute for Experimental Therapy. In 1908 Ehrlich shared the Nobel Prize in physiology or medicine with Élie Metchnikoff for their separate paths to an understanding of the immune response: Ehrlich presented a chemical theory to explain the formation of antitoxins, or antibodies, to fight the toxins released by the bacteria, while Metchnikoff studied the role of white blood corpuscles in destroying bacteria themselves. By that time most scientists agreed that both explanations of the immune system were necessary.

Early in his career Ehrlich began to develop a chemical structure theory to explain the immune response. He saw toxins and antitoxins as chemical substances at a time when little was known about their exact nature. Up to that time those scientists, like Felix Hoffmann (p. 10), who were synthesizing therapeutic agents came at their tasks with few hypotheses about where and how these agents interacted with living systems. Ehrlich supposed that living cells have side chains much in the way that dye molecules were known to have side chains that were related to their coloring properties. These side chains can link with particular toxins, just as the organic chemist Emil Fischer had said that enzymes must bind to their receptors "like a key in a lock." According to Ehrlich, a cell under threat from foreign bodies grows more side chains, more than are necessary to lock in foreign bodies in its immediate vicinity. These "extra" side chains break off to become antibodies and circulate throughout the body. It was these antibodies, in search of toxins, that Ehrlich first described as magic bullets. Serum therapy was for him the ideal method of contending with infectious diseases. In

■ Paul Ehrlich. Courtesy Edgar Fahs Smith Collection, University of Pennsylvania Library.

those cases in which effective sera could not be discovered, Ehrlich would turn to synthesizing new chemicals, informed by his theory that the effectiveness of a therapeutic agent depended on its side chains. These chemotherapies were to be the new magic bullets.

Over time Ehrlich elaborated his theory of the immune system, supposing, for example, that a disease-causing substance must have special chemical groups that enable it to link to cells and others that account for the toxic action of a substance. Sometimes in explaining his theories, Ehrlich became so excited that he wrote on virtually any available surface, covering tablecloths, shirt cuffs, picture postcards, and laboratory walls with diagrams.

In Frankfurt, Ehrlich returned from his work on sera to chemotherapies and dyes. First targeting trypanosomes—the protozoa that were known to be responsible for certain diseases such as sleeping sickness—he and the Japanese bacteriologist Kiyoshi Shiga synthesized trypan red as a highly effective cure for that disease.

In 1906 Georg-Speyer-Haus, a research institute for chemotherapy, was established with its own staff under Ehrlich's direction. Soon this institute and the Hoechst and Cassella chemical companies reached an agreement that gave the companies the right to patent, manufacture, and market preparations discovered by Ehrlich and his colleagues. The companies further agreed to supply chemical intermediates for the syntheses that the staff of the institute would undertake.

The researchers, now including an organic chemist, Alfred Bertheim, and a bacteriologist, Sahashiro Hata, broadened the targeted microorganisms to include spirochetes, which had recently been identified as the cause of syphilis. Beginning with an arsenic compound, atoxyl, in three years' time and three hundred syntheses later—for that day an amazingly large number—they discovered salvarsan (1909), or dihydroxydiaminoarsenobenzenedihydrochloride (see introduction to chapter 5, p. 64). Salvarsan was first tried on rabbits that had been infected with syphilis and then on patients with the dementia associated with the final stages of the disease. Astonishingly, several of these "terminal" patients recovered after treatment with salvarsan. More testing revealed that salvarsan was actually more successful if administered during the early stages of the disease. Salvarsan and Neosalvarsan (1912) retained their role as the most effective drugs for treating syphilis until the advent of antibiotics in the 1940s.

# 2. Early Pharmaceutical Companies and the Pure Food and Drugs Act

The Pure Food and Drugs Act was passed in 1906 in response to a long-standing concern about food and drug adulteration. Before that time virtually no government regulation existed for ensuring the purity or quality of pharmaceutical products, and manufacturers frequently adulterated their medicines in an effort to increase their profits. These adulterations not only cheated the public but in many cases also threatened the public's health.

In contrast, some early pharmaceutical companies—like George K. Smith and Company (now GlaxoSmithKline) and E. R. Squibb and Company (now Bristol-Myers Squibb)—objected to shoddy quality in pharmaceuticals. They instituted strict standards for quality and purity, and as a result their products became known and trusted by the public.

During the nineteenth century public concern over adulteration increased, which caused a few pieces of legislation to be passed, on both the state and the federal levels. But enforcement was lax, and adulteration continued. When the Pure Food and Drugs Act was finally passed in 1906, it was the most comprehensive anti-adulteration legislation to date, a landmark bill that required the cooperation of many dedicated individuals, pharmaceutical companies, and activist groups.

## George K. Smith (1810–1864)

The ideals that led to the passage of the Pure Food and Drugs Act of 1906 had a long history in the pharmaceutical industry. George K. Smith, brother and longtime partner of John K. Smith, had always emphasized pure products and guaranteed quality.

Born to Johannes Klein Schmidt and Catharina Schmidt in 1810, George was one of seven sons. He grew up in Upper Salford, Pennsylvania, and was probably apprenticed to an apothecary as a teenager. At the age of twenty-one George received his share of his father's estate (his father died when George was only thirteen years old), and a year later he attended the Philadelphia College of Pharmacy. Like his older brother John, he chose pharmacy as his profession. In 1841 George joined John's company, John K. Smith and Company, at 296 North Second Street in Philadelphia.

At the time government did not regulate the purity or quality of pharmaceutical products; thus manufacturers frequently adulterated their medicines in an effort to increase their profits. Unlike many pharmacists of the time, Smith quickly recognized the importance of purity in his products. His remarkably high standards combined with his efficient running of the business earned him an excellent reputation in Philadelphia and the surrounding area.

After the death of his brother in 1845 Smith continued the business as George K. Smith and Company, which quickly became an extremely successful pharmaceutical wholesaler. His was the first drug house to cater to the precise requirements of physicians, making him particularly popular among the medical community, and he used manufacturing methods that were quite advanced for his time.

Given his strong belief in the quality of his products, Smith was perhaps the company's best salesman, and he frequently made sales trips to central Pennsylvania and New Jersey, collecting orders. These trips paid off: his reputation grew, and dealers from the outlying areas would make annual or semiannual trips to his store, purchasing enough to last them for at least six months. Business increased further with the advent of the Mexican-American War in 1846; George Smith supplied quinine and other medications to the U.S. Army.

In 1851 Smith's prosperity enabled him to purchase a property adjoining 296 North Second Street. Four years later he further expanded by opening a second store just a few blocks away. By that time the business was large enough for him to admit two partners: E. L. Trimble and Mahlon K. Smith, George's nephew.

Though the Civil War was hard on Smith's business, by 1868 sales had risen again. His reputation for honesty—both in his business and his personal life—aided him in garnering financial support from family and friends. Unfortunately, Smith himself did not live to see his business flourish again; he died in 1864 at the age of fifty-four. His company's commitment to quality products did continue, however, as Mahlon K. Smith took over.

Today Smith's company still exists. Now called GlaxoSmithKline, it is a well-known producer of pharmaceuticals worldwide.

■ George K. Smith. Courtesy SmithKline Beecham.

# EDWARD ROBINSON SQUIBB (1819–1900)

Edward Robinson Squibb devoted his career to the struggle for purity and consistency in medicines. He was born on 4 or 5 July 1819, in Wilmington, Delaware, to James Squibb and Catherine Harrison Bonsall Squibb, who were devout Quakers. In 1831, when Edward was twelve, his mother and his three sisters died. As a result Squibb was sent to live with his Grandmother Bonsall in Darby, Pennsylvania. Several years later Squibb's father suffered a stroke that left him an invalid for the rest of his life.

Squibb remained with Grandmother Bonsall until he was eighteen, when he began his apprenticeship with the pharmacist Warder Morris in Philadelphia. Later he worked for another pharmacist, J. H. Sprague. In 1842 Squibb fulfilled his dream of studying medicine, entering Jefferson Medical College. He emerged three years later with an M.D.

Soon after completing his studies, Squibb began to question his Quaker beliefs. The Mexican-American War had broken out, and Squibb's medical abilities were sorely needed by the U.S. armed forces. Ultimately, his conscience decided the matter: in 1847 Squibb became an assistant surgeon with the U.S. Navy, an appointment that led his Quaker meeting to disown him.

In his capacity as a commissioned assistant surgeon, Squibb began his crusade against impure pharmaceuticals. While sailing on the *Cumberland*, Squibb took it upon himself to toss large quantities of inferior medications overboard.

As his medical career with the navy progressed, Squibb's commitment to improving the quality of drugs increased. In 1852 he was assigned to the Brooklyn Naval Hospital, where he planned to work with Benjamin Franklin Bache. Although Congress had passed and President James Polk had six years earlier signed a measure that prohibited the import of impure pharmaceuticals, it was not enforced, and the problems remained. Bache and Squibb intended to manufacture pure drugs at the Brooklyn Naval Hospital, for use at the hospital and on the ships it supplied.

Once his laboratory at the Brooklyn Naval Hospital was set up, Squibb began producing various substances, including ammonia, blue pill (a mercurial used to treat syphilis), potassium iodide, citric acid, tincture of opium, and ether, a new anesthetic. In 1853 Squibb received a visit from James C. Dobbin, Secretary of the Navy under President Franklin Pierce. Dobbin toured the new laboratory at Brooklyn Naval Hospital and, approving of the work in progress, promised to secure funds for the lab. In addition, Squibb was removed from his general hospital duties and ordered to spend all his time in the laboratory.

After several years Squibb decided to leave the navy for private industry. Although the products that left his laboratory met Squibb's strict standards for purity and quality, he was often frustrated by the poor quality of the materials he received from suppliers. In 1857 he signed a one-year agreement with Laurence Smith, a chemist, and Thomas E. Jenkins, the owner of a Louisville, Kentucky, drugstore, to join their commercial laboratory, known as the Louisville Chemical Works. Squibb ran the laboratory for his partners, an experience that solidified his desire to begin his own pharmaceutical business.

Once the contract with the Louisville Chemical Works was fulfilled, Squibb returned to Brooklyn to set up his own manufacturing company. The U.S. Army had indicated that it would order the bulk of its drugs from him once he was open for business. Taking this promise to heart, Squibb began production in December 1858. Just a few weeks later, however, a broken bottle of ether caused a fire in his laboratory. His building suffered extensive damage, and Squibb himself was burned in an attempt to save his ledgers and scientific notes from the blaze. Although the laboratory was rebuilt, Squibb bore the scars of this fire for the rest of his life.

Aside from his business, to which he devoted innumerable hours, Squibb also became active in many pharmaceutical and medical organizations. These professional organizations provided Squibb with forums to effect the changes he felt necessary to improve the quality of pharmaceuticals. The first opportunity to do so on a national level came when he was selected for the 1860 Committee for the Revision of the U.S. Pharmacopoeia, a register of approved drugs, chemicals, and medicinal preparations that included tests for the purity and strength of such products. Into this private, nonprofit committee Squibb introduced his strict standards for quality, resulting in a pharmacopoeia that more closely matched his ideal.

Squibb's ideals, however, were not always readily accepted.

■ Edward Squibb. Courtesy Bristol-Myers Squibb Corporation.

He felt that pharmacy should be more closely connected with medicine—as in his own life—and was frustrated by the preponderance of pharmacists, often patent profiteers or pseudo-chemists, at professional meetings. This conflict led Squibb to reject the invitation to join the 1870 Committee for the Revision of the U.S. Pharmacopoeia. In addition, at the 1876 Convention of the American Medical Association, Squibb moved to put the pharmacopoeia under the control of physicians rather than pharmacists. This proposal was discussed at the 1877 convention as well but was defeated.

Squibb continued to work toward legislation that restricted adulteration in pharmaceuticals. He joined a committee of New York State groups organized to discuss options for such legislation. The report of the committee to the 1879 Convention of the New York State Medical Society was readily supported, and their suggestions were enacted into law by both New York and New Jersey. The measure also was introduced into the U.S. Congress, but to Squibb's disappointment it died in committee.

Although that was Squibb's final political effort to protect against adulteration, he continued to publicize his views on the subject. Beginning in 1881, he published *An Ephemeris*, in which he discussed various issues concerning the adulteration of substances as well as the disjuncture between pharmacy and medicine. Even though his commentaries frequently resulted in harsh debate, Squibb continued this publication with few interruptions until his death in 1900.

# HARVEY WASHINGTON WILEY (1844–1930)

A pure food and drug bill would probably not have passed in 1906 without the unceasing commitment of one man: Harvey Washington Wiley. A doctor and a chemist, he recognized early the need to regulate the adulteration of food and later joined the crusade for pure drugs.

Wiley was born on 14 October 1844, on the family farm near Kent, Indiana. Until 1863, when he decided to enroll at Hanover College, he (like his parents) had little formal education—only five winters of instruction.

In 1865, after serving in the Civil War, Wiley returned to Hanover to complete his studies. Shortly after graduating, he accepted a position teaching in a Lowell, Indiana, public school and spent his spare time learning about medicine from a local doctor. Four years later Wiley enrolled at Indiana Medical College and after two terms earned his medical degree.

Rather than going into private practice, Wiley continued to teach, first at the secondary level in Indianapolis and later at Indiana Medical College and Northwestern Christian University. Wiley also studied for a year at the Lawrence Scientific School of Harvard University, emerging with a B.S. in science. Afterward he resumed teaching his classes at Northwestern Christian University as a professor of both chemistry and mineralogy. In 1874 Wiley took a position as a professor of chemistry at Purdue University.

Wiley taught for many years at Purdue but began to expand his concerns beyond academia. Beginning in 1880, he protested against the practice of adulteration in fertilizers and in food. In 1881 Indiana regulated the manufacture and sale of commercial fertilizers, and Wiley was appointed state chemist to analyze the contents of these products. Two years later he was appointed chemist of the U.S. Department of Agriculture (USDA) Chemical Division.

Wiley's primary job at the USDA entailed crystallizing sugar from sorghum cane as an alternative to sugar cane. Wiley spent much of his time on the sugar experiments, but shortly after arriving at the USDA, he allocated funds to developing analytical methods and investigating food adulteration. In 1884 and 1885 he conducted studies on milk, butter, and honey. Two years later the department began publishing Technical Bulletin 13, *Foods and Food Adulterants*. For the next several years issues of the bulletin addressed the adulteration of such products as dairy foods,

spices, and alcoholic beverages. Recognizing the need for broader public awareness of these issues, in 1890 the department issued Bulletin 25, *A Popular Treatise on the Extent and Character of Food Adulterations*, which advocated national legislation.

With public concern over adulteration growing, commercial interests voiced their support of legislation. In response the 1887 and 1888 national conventions of the American Society for the Prevention of the Adulteration of Food attempted to develop appropriate legislation. Congress passed legislation in 1888 that prevented the sale or manufacture of adulterated food or drugs in the District of Columbia. In 1890 and 1891 legislation addressing the inspection of meat was passed. The 1890 law also prohibited the importation of adulterated foods, drugs, or beverages. Unfortunately, enforcement was lax, making a stronger bill necessary.

When the sugar experiments ended in 1895, so did the food adulteration work by the USDA. However, six years later, in 1901, the area over which Wiley presided—the USDA's Division of Chemistry—became the Bureau of Chemistry. This change led to the establishment of a food laboratory.

Although a few legislative successes had been won in the early 1890s, Wiley persisted. In 1895 he chaired a committee of the Association of Official Agricultural Chemists (AOAC) to develop a national anti-adulteration law. The AOAC was originally formed to address the adulteration of fertilizer, but under Wiley's direction it expanded its interests to the adulteration of food and drugs. In 1897 the committee made its recommendation. The bill it put forward dealt primarily with ensuring honest labeling but also prohibited the use of any harmful adulterant. A number of groups supported this effort, including the National Pure Food and Drug Congress held in 1898. Wiley presented a key address at this meeting and chaired the legislation committee considering the proposed bill. Though a revised version of this bill was brought to Congress in 1899, no action was taken. The bill was reintroduced the following year, along with several other less comprehensive versions of pure-food legislation.

Debate continued over the proposed legislation for several years with little result. But in 1902 the USDA was authorized to establish food standards. Although the research

■ Harvey Washington Wiley. Courtesy FDA History Office, Rockville, Maryland.

required was largely done by the AOAC, Wiley retained his influence and consulted not only manufacturers but also consumers. The year 1903 brought yet another duty for the USDA: the control of food imports. As the efforts to bring about far-reaching legislation continued after 1903, studies conducted by Wiley on the effects of preservatives on the human body became very influential. The first results of these experiments were published in 1904. Newspapers also carried sensationalist articles about the studies, dubbing the experiment's volunteers the "Poison Squad" and thus capturing the interest of the public.

Although his crusade focused primarily on food adulteration, Wiley's interests eventually widened into drug adulteration. Wiley spoke on the issue as early as 1903, and when the American Medical Association and the American Pharmaceutical Association banded together in 1905 to form a Council on Pharmacy and Chemistry to combat the adulteration of pharmaceuticals, Wiley joined.

Wiley was an unceasing advocate of national pure food and drug legislation and considered it his life's work. While debates in Congress continued for years, Wiley remained in the background, providing information and statistics, lecturing across the nation about the perils of adulteration, and organizing groups—particularly women's groups—to support the legislation. Finally, with the backing of President Theodore Roosevelt, the landmark Pure Food and Drugs Act was passed by Congress in 1906 and signed into law. The bill prohibited adulteration and misbranding of food and drugs and was the most comprehensive legislation of its kind in the United States.

After the bill was passed, Wiley continued his advocacy of the issue, emphasizing the need for strict standards in product labeling. After leaving the USDA in 1912, he became director of the *Good Housekeeping* Bureau of Foods, Sanitation, and Health.

# 3. Vitamins

Today, vitamins come in all shapes and sizes: they are chewable or in caplet form, are often mixed with minerals to provide your "daily supplement," and artificially "fortify" many of our foods. From A to K, vitamins are necessary players in animal metabolism, often acting as coenzymes in catalytic reactions.

Before the twentieth century the only source of vitamins was food. Diseases of vitamin deficiency were common, especially in places where diets were restricted, such as on ships. For centuries sailors suffered from a variety of debilitating diseases, including scurvy and beriberi, which are now known to be linked to vitamin C and vitamin D deficiencies, respectively. Physicians investigating the health and diet of sailors found they could control and even prevent the diseases through the prescription of certain foods, but a category of specific disease-combating factors in foods was not identified until the early twentieth century.

While investigating beriberi in 1911, Casimir Funk, a chemist at the Lister Institute in London, crystallized an amine substance from rice bran. He knew that something in rice bran prevented beriberi, and convinced he had found that substance, he named his discovery "vitamine," for vital amine. In 1920 the *e* was dropped from *vitamine*, as the *ine* mistakenly connoted that all such substances had an amine nature. Thus the word *vitamin* entered our language.

Vitamins are now mass-marketed in pill form. The synthetic production of vitamins came about through a multi-step process of discovery and research. Once the existence of a vitamin was hypothesized, scientists often spent many years attempting to isolate the vitamin from a food source in which it was abundant. Isolating the vitamin then made it possible to develop a process for synthesizing it in the lab. Often, different researchers carried out different steps in the process, so it is difficult to credit one individual for the bottle of vitamin C or E on the drugstore shelf. (Besides the biographies included in this section, see also Tadeus Reichstein on vitamin C and Max Tishler on vitamin $B_2$ [p. 97] and Robert Burns Woodward on vitamin $B_{12}$ [p. 114]).

# ALBERT SZENT-GYÖRGYI (1893–1986)

In 1769 William Stark, a young British physician, began a series of experiments on diet and nutrition. Unfortunately, he used himself as the experimental subject. After consuming only bread and water for thirty-one days, Stark added other foods to his diet, one by one, including olive oil, figs, goose meat, and milk. In two months Stark recorded that his gums were red and swollen, and they bled when touched. Seven months later he died, probably from scurvy, definitely as a result of malnutrition. His diet, heavy on meat and starch and devoid of fresh vegetables and citrus fruits, resembled that of British sailors. Scurvy had debilitated hundreds of thousands of sailors and other people by the mid-eighteenth century. Ironically, the first step in eradicating the disease came in 1757, twelve years before Stark's ill-fated experiment. James Lind, a Scottish physician, wrote a treatise recommending mandatory consumption of citrus fruits and lemon juice by sailors in the British Navy. During his years as a British naval surgeon Lind observed the curative and preventive powers of citrus fruits in sailors suffering from scurvy. It was not until 1932, however, that the link between vitamin C and the treatment and prevention of scurvy was explicitly understood.

In 1907, after Axel Holst and Alfred Fröhlich posited the existence of vitamin C on the basis of its biological effects, an international competition ensued to isolate the vitamin. In 1928 Albert Szent-Györgyi, a Hungarian-born physician and biochemist, isolated a substance that was identified four years later as vitamin C. Oddly enough, Szent-Györgyi's main scientific interest was not vitamins but rather the chemistry of cellular metabolism.

Born into a family that included three generations of scientists, Szent-Györgyi was inclined toward science from an early age. He enrolled in the university in Budapest in 1911 to study medicine, but his education was interrupted by the outbreak of World War I. Fervently antiwar throughout his life, Szent-Györgyi wounded himself in order to escape combat service after a short time in the army and returned to the university in 1917. Reflecting on the experience, he said, "I was overcome with such a mad desire to return to science that one day I grabbed my revolver and in my despair put a shot through my upper arm."

Szent-Györgyi received his medical degree the year he returned to Budapest; he went on to study at various other European universities. He was interested in cell respiration and the production of energy, and during this time he investigated the "browning" process in plants. He discovered that plants that turn brown upon withering do so as a result of damaged reducing mechanisms at the cellular level; a malfunctioning reducing mechanism cannot supply enough hydrogen to prevent oxidation. Szent-Györgyi then turned to plants that do not turn brown when withering. While carrying out a series of experiments on citrus plants, Szent-Györgyi found that he could induce browning with peroxidase—a plant enzyme active in oxidation—and then delay that browning with the addition of citrus juice to the peroxidase. Szent-Györgyi isolated the agent in the citrus juice responsible for countering browning, naming it "hexuronic acid."

For two years Szent-Györgyi searched for an ample source of hexuronic acid to continue his experiments. He spent time at Cambridge University and then at the Mayo Foundation in Minnesota, all the while researching his discovery. In 1930 he returned to Hungary and took up a position as professor of medicinal chemistry at the University of Szeged. In Szeged, Szent-Györgyi showed his sample of hexuronic acid to J. L. Svirbely, an American-born chemist of Hungarian descent, who had previously worked with Charles G. King, a vitamin researcher at the University of Pittsburgh. Svirbely performed the classic feeding experimentation on guinea pigs, which like humans, are unable to produce their own vitamin C. He fed half the guinea pigs with boiled food—boiling already being known to destroy vitamin C—and half with food enriched with hexuronic acid. While the one group developed scurvy-like symptoms and died, the other group flourished. It was clear to Szent-Györgyi and Svirbely that hexuronic acid was indeed vitamin C. Meanwhile, through a still-disputed series of events and communications, King came to the same conclusion and published an announcement in *Science* magazine two weeks before Szent-Györgyi's note appeared in *Nature*.

After renaming hexuronic acid "ascorbic acid" to reflect its anti-scurvy properties, Szent-Györgyi began to search for abundant natural sources of the acid. Szeged was then and is now a major center of paprika cultivation, and—according to one version of the legend—to avoid eating the fresh paprikas that his wife had served as a side dish, Szent-

■ Albert Szent-Györgyi. Courtesy Edgar Fahs Smith
Collection, University of Pennsylvania Library.

Györgyi hurriedly left the dinner table and ran to his laboratory to test the paprikas for ascorbic acid content. Eureka! With an ample source of ascorbic acid identified, work continued on vitamin C, and within two years its structure was known and it was being synthesized in the laboratory by Szent-Györgyi's collaborator, Walter Haworth, at the University of Birmingham in England. Moreover, its therapeutic effects—one of which was the prevention of scurvy—were under extensive investigation.

Vitamin C figured prominently in the Nobel Prizes in science in 1937. Szent-Györgyi received the prize in physiology or medicine for his discoveries concerning biological combustion, including the role of vitamin C in the process. For his carbohydrate researches and his structural determination of vitamin C, Walter Haworth shared the prize in chemistry with the Swiss chemist Paul Karrer, who had also done some work on vitamin C but was more famous for research on vitamins A and $B_2$.

Szent-Györgyi went on to study muscle contraction; he identified actin and myosin, the proteins responsible for the physiological process of contraction. He later carried out further studies of citrus fruits, identifying vitamin P (a complex compound of flavenoids) and postulating its use in strengthening capillaries. A deficiency disease linked to the lack of vitamin P, however, has never been identified.

Szent-Györgyi, who became a passionate advocate of government-sponsored cancer research after losing two close family members to the disease, spent much of his later professional life studying cancer at the cellular level, at the Woods Hole Institute for Muscular Research. An outspoken opponent of military spending, nuclear weapons, and war in general, Szent-Györgyi was pessimistic about the state of modern life and expressed his sociopolitical views in his 1970 book, *The Crazy Ape*.

An active researcher until World War II, Charles King subsequently cultivated the scientific work of others by teaching and providing funds to researchers. He was a Columbia University professor from 1941 to 1962 and led the Nutrition Foundation for many years. Svirbely led a quiet life in research, ultimately working at the headquarters of the U.S. Public Health Service.

# ROBERT R. WILLIAMS (1886–1965)

Almost 150 years after Scottish naval surgeon James Lind prescribed citrus fruit to British sailors to curtail scurvy, another naval doctor, Kanehiro Takaki, of the Japanese Naval Medical Services, ordered Japanese sailors to increase their intake of vegetables, barley, fish, and meat. Noting that polished white rice made up the bulk of sailors' rations and that there was a high incidence of beriberi among those sailors, Takaki made the link between this debilitating disease and the diet of the sailors. Within a decade of Takaki's order, beriberi had been all but eliminated in the ranks of the Japanese Navy, lending further credibility to Takaki's hypothesis.

During the same period the Dutch physician Christiaan Eijkman was conducting his own research in Java on the cause of beriberi. Although erroneously convinced throughout his research career that bacteria caused the disease, Eijkman made a valuable contribution to the study of beriberi. He conducted a series of experiments with laboratory chickens, demonstrating that beriberi could be induced if the animals were fed a diet consisting exclusively of polished rice.

A few islands over, in Manila, Robert Williams, an American chemist with the Bureau of Science, was enlisted to research the theory that beriberi was caused by a nutritional deficiency. Williams, the son of a missionary, was born in India; after earning his master's degree in chemistry from the University of Chicago, he decided to return to Southeast Asia and found employment with the Bureau of Science in the Philippines. Unable to ascertain the substance in unpolished rice that prevented beriberi, Williams came back to the United States.

He did not, however, give up his research. In 1925, while employed at Bell Labs, Williams began to experiment in his garage, using his wife's washing machine as a centrifuge and his own money to support his research. One year later two Dutch chemists, Barend Jansen and Willem Donath, working in Eijkman's laboratory, crystallized the antiberiberi factor (vitamin $B_1$), which they called "aneurin," from rice bran. Jansen and Donath were incorrect in the formula they deduced for the vitamin, however; they missed the sulfur atom, thereby making synthetic production impossible.

Williams continued researching vitamin $B_1$. In the early 1930s he approached Merck and Company for funding and support of his work on the vitamin. Merck agreed to supply the needed crystalline vitamin and also provided lab space and assistants to Williams. The investment paid off shortly thereafter; in 1936 Williams made the correct structural determination of vitamin $B_1$ and designed a synthesis for it. Williams named vitamin $B_1$ "thiamin" and submitted it for addition to the American Medical Association's publication *New and Non-Official Remedies*. The American Chemical Society added an *e* to the name to reflect the amine nature of the vitamin. Thiamine, found in whole-grain cereals, meats, yeast, and nuts, acts as a cofactor in the enzymatic reaction that breaks down carbohydrates, alcohol, and some proteins.

■ Robert R. Williams in 1938. Courtesy Merck Archives, Merck & Co., Inc.

■ Robert Williams, demonstrating the structural formula of vitamin B$_1$. Courtesy Merck Archives, Merck & Co., Inc.

In 1947 Williams was awarded the Perkin Medal by the Society of Chemical Industry, recognizing his determination of the structure of vitamin B$_1$ and development of a commercially viable synthesis for the compound. Williams donated all his patent royalties from the synthesis of the vitamin to the Williams-Waterman Fund. This fund was administered through the Research Corporation, a philanthropic foundation dedicated to funding scientific research and active in combating diseases of malnutrition in underdeveloped countries of the Western Hemisphere.

## Gladys Ludwina Anderson Emerson (1903–1984)

In 1927 twenty-three-year-old Gladys Anderson Emerson, a Stanford-trained economic historian, accepted a fellowship in nutrition at the University of California at Berkeley, thus fortuitously changing the course of her professional career. A young woman with many interests and multiple career paths open to her, Emerson chose animal nutrition and biochemistry. Her decision brought her into a research team whose work resulted in the isolation of vitamin E.

Born on 1 July 1903 in Caldwell, Kansas, Gladys Anderson was the only child of parents of Swedish descent. While their daughter was still an infant, the Andersons moved to Fort Worth, Texas; twelve years later they relocated to El Reno, Oklahoma. Young Gladys was an avid student, showing simultaneous promise in mathematics, history, Latin, chemistry, and music. After graduating from high school, she enrolled in the Oklahoma College for Women (now the University of Sciences and Arts of Oklahoma), intending to become a teacher.

Emerson was subsequently offered assistantships in both the chemistry and the history departments at Stanford University; she chose history. She received her master's degree in history, with a minor in economics, from Stanford a year later, in 1926. However, while working toward her master's degree, Emerson also managed to work in courses in physical chemistry. Never forgetting her interest in the sciences, Emerson taught at the junior-high level in Oklahoma for a few years but jumped at the chance of a fellowship in nutrition and biochemistry when it was offered to her by the University of California at Berkeley.

Five years later she received her doctorate in animal nutrition and biochemistry. After a postdoctoral year at the University of Göttingen, where she worked with Nobel laureates Adolf Windaus and Adolf Butenandt, Emerson returned to Berkeley. She took a position in the lab of Herbert M. Evans, director of the University of California's Institute of Experimental Biology.

For three years Emerson worked with Evans on isolating vitamin E in pure form from its natural sources. Evans had first identified and named vitamin E in 1922 but had been unsuccessful in isolating it. In 1936 the team successfully isolated from wheat germ oil a pure form of vitamin E, which they named "tocopherol." Emerson and Evans went on to identify two different forms in which the vitamin could be isolated, alpha-tocopherol and beta-tocopherol. Their research paved the way for the subsequent determination of the chemical structure of tocopherol, which made the later synthesis of artificial vitamin E possible.

That vitamin-E deficiency affected levels of fertility in laboratory animals had been known for years before Emerson began her research. While at Berkeley, Emerson conducted research that further strengthened the connection between vitamin E and fertility. In her experiments four generations

■ Gladys Emerson in the early 1950s. Courtesy Merck Archives, Merck & Co., Inc.

■ Gladys Emerson with her camera and dog in 1952.
Courtesy Merck Archives, Merck & Co., Inc.

of lab rats were given less and less vitamin E, with the result of increasing infertility; yet fertility could be restored in the fourth generation via dosages of vitamin E. She also went on to demonstrate in lab rabbits that, besides affecting fertility, a controlled dietary deprivation of vitamin E could cause a reaction akin to muscular dystrophy.

When Emerson published the results from her research on vitamin E, she received a great deal of recognition within the scientific community. In 1942 she was invited to join the Merck Institute for Therapeutic Research, in Rahway, New Jersey. Making the transition from academia to industry, Emerson continued her work on the effects of vitamins: she began a research program on the whole B complex of vitamins. Emerson proved the link between vitamin B–deficient diets and abnormalities of growth and posture and of the eyes, skin, liver, kidneys, and other internal organs. Not only did Emerson prove this relationship, but she also devoted her research efforts to finding more effective methods of delivering the vitamins to reverse the negative effects of deficiency—injection instead of oral administration, for example.

In 1956 the University of California at Los Angeles invited Emerson to join their faculty as a professor of nutrition and the chairperson of the department of home economics. Emerson accepted, and while assuming a teaching role, she continued as an active researcher in her field. By 1962 she had also been appointed the vice chair of UCLA's department of public health.

Among many other awards Emerson received the Garvan Medal in 1952. This award, given by the American Chemical Society, recognizes "distinguished service to chemistry by women chemists." Emerson was also appointed as vice chair of the Panel on the Provision of Food in 1969 by President Richard Nixon, and in 1970 she served as an expert witness before the U.S. Food and Drug Administration in the Hearings on Vitamins and Mineral Supplements and Additives to Food.

■ Carl Dam in his laboratory holding a syringe and a test tube. Courtesy Edward G. Miner Library, University of Rochester Medical Center, New York.

## Carl Peter Hendrik Dam (1895–1976), Edward Adelbert Doisy (1893–1986), and Karl Paul Gerhardt Link (1901–1978)

The discovery of vitamin K and its role in blood clotting laid the foundation for the development—twenty years later—of the blood thinner Coumadin (warfarin). Although initially developed as a rat poison, warfarin is today's most-prescribed medication for medical conditions involving blood clots.

In the beginning of the twentieth century, farmers in the northern prairie states began to plant sweet clover. Standard feed crops would no longer grow in their over-farmed land, so sweet clover, imported from Europe, was required to feed the cattle. Unfortunately, in the 1920s, a new disease was reported, one that decimated cattle herds. "Sweet clover disease," as this new malady was named, was traced to spoiled sweet clover hay. Cattle succumbing to the disease developed spontaneous and fatal internal bleeding. Their blood would not clot, so the disease could be cured only by blood transfusions from healthy cattle.

Scientists discovered in 1929 that cattle with sweet clover disease had great deficiencies of prothrombin, a plasma protein that, it was later found, is produced in the presence of vitamin K and converted into thrombin, which facilitates the clotting of blood. Vitamin K itself was identified in the 1930s. Early in that decade Danish chemist Carl Peter Hendrik Dam conducted a series of animal studies that demonstrated that an unknown vitamin was necessary for blood coagulation. In 1935 he named this unknown substance "vitamin K."

Dam was born in 1895 in Copenhagen. His father was a pharmaceutical chemist, and Dam followed in his footsteps, receiving his M.S. in chemistry in 1920 from the Polytechnic Institute. He taught chemistry at the Royal School of Agriculture and Veterinary Medicine before joining the University of Copenhagen, where he studied physiology. In 1928 he became an assistant professor of biochemistry at the University of Copenhagen.

Soon after Dam named vitamin K, Edward Adelbert Doisy, an American biochemist working at the School of Medicine at the University of St. Louis, discovered that Dam's vitamin K was actually two compounds: vitamin $K_1$—produced by plants—and vitamin $K_2$—the form to which all vitamin-K compounds are converted in the human body.

Doisy was born in Hume, Illinois, in 1893. He studied a premedical curriculum at the University of Illinois, where he received his B.A. in 1914 and his M.S. in 1916. He continued his graduate study at Harvard, receiving his Ph.D. in 1920. In 1919, not having quite finished his doctorate, Doisy joined Washington University as an instructor in biochemistry. Four years later he became head of the department of biochemistry at the University of St. Louis School of Medicine.

In 1938 Doisy and his colleagues determined the structures of the two vitamin-K compounds and a year later synthesized vitamin $K_1$. For this work Doisy and Dam were awarded the 1943 Nobel Prize in physiology or medicine.

The mystery behind sweet clover disease was solved by Karl Paul Link, an agriculturalist employed at the University of Wisconsin's Agricultural Experiment Station in Madison. Link was born in 1901 in La Porte, Indiana. The son of a Lutheran minister, Link showed an aptitude for chemistry during high school. When his teacher was called to serve in World War I, the high school selected Link to teach the chemistry classes. Link attended the University of Wisconsin, earning his B.S. in 1922 and his Ph.D. in 1925.

Although Link was hired by the Agricultural Experiment Station to develop a strain of sweet clover free from coumarin—the compound responsible for sweet clover's bitter taste—in 1933 he decided to pursue the problem of sweet

■ **Edward Doisy. Courtesy Saint Louis University Archives.**

39

■ Karl Link preparing a cocktail of Warficide, one of the trade names under which warfarin is sold as rat poison. Courtesy University of Wisconsin—Madison Archives.

clover disease. This decision was undoubtedly inspired by his encounter with a farmer whose cattle were killed by the deadly disease. The farmer came to him in the middle of a blizzard—with a dead cow, spoiled sweet clover, and a bucket of non-clotting blood in hand—asking for help. Link could only recommend the expensive treatment of blood transfusions for the sick cattle, but he vowed to solve the problem of sweet clover disease. Within six years Link and his colleagues identified the hemorrhagic agent of sweet clover disease and determined its structure. This hemorrhagic agent —3,3′-methylene bis-(4-hydroxycoumarin)—was later named "dicumarol." Dicumarol is formed naturally when coumarin oxidizes and combines with other substances present in moldy hay. Link also demonstrated that vitamin K reversed the hemorrhagic effects of dicumarol.

In 1941 dicumarol was shown to work as an anticoagulant in humans. In the same year the Wisconsin Alumni Research Foundation—to which Link assigned his patent rights—applied for a patent for dicumarol. The foundation then licensed Eli Lilly, Squibb, and Abbott Laboratories to produce the drug. By 1944 dicumarol was widely available and used to treat thrombosis—a potentially fatal condition caused by the formation of a blood clot in a blood vessel or in the heart—and myocardial infarction, or heart attack.

In 1945 Link developed tuberculosis and was confined to a sanatorium. While there, he began to read about the history of rat poison. When he returned to work in 1946, Link wanted to create the ideal rat poison. Several years earlier he had tried dicumarol—because of its ability to induce fatal internal bleeding—as a rat poison, but he found it unreliable. In the course of his research on the compound, however, his laboratory synthesized more than a hundred variations of dicumarol. Link found that several of these compounds were good rodenticides, and in 1948 he introduced 3-phenylacetyl ethyl, 4-hydroxycoumarin as warfarin. Link named this compound after the Wisconsin Alumni Research Foundation.

Two years later, when dicumarol faced increasing competition from European drugs, Link tried to convince clinicians to try warfarin in humans. When a navy recruit unsuccessfully tried to commit suicide with 567 milligrams of warfarin in 1951, the clinicians agreed to test the compound. By 1953 they determined that warfarin was much more effective than dicumarol, and Endo Products, Inc., introduced it commercially a year later as Coumadin. (Endo was subsequently acquired by DuPont, whose pharmaceutical division is now part of Bristol-Myers Squibb.) Coumadin soon became a standard treatment for long-term treatment of thrombotic conditions.

Despite this success scientists did not understand how warfarin works until the 1970s. Warfarin inhibits the enzyme vitamin K epoxide reductase, thereby blocking the action of vitamin K and prohibiting proper blood clotting.

# 4. Enzymes, Hormones, and Neurotransmitters

In the early twentieth century, scientists made great strides in understanding the roles of the body's own chemicals in regulating its functions and transmitting information among its parts. Among these chemical substances were enzymes, hormones, neurotransmitters, and histamines—all investigated successfully in the first three decades of the twentieth century.

Much of the work was really directed toward obtaining a basic biological and chemical understanding of the identity and function of these body substances, but the pharmaceutical implications of the investigations were near at hand. Since many disease conditions had long been thought to be caused by some sort of imbalance in the patient's own body, it seemed obvious to twentieth-century researchers that assisting or interfering with natural regulators and transmitters ought to provide treatments for many maladies. In most areas, though, the applications would come much later.

Replacement therapy, which is needed when the patient's body cannot supply a necessary substance in sufficient quantities, was the most obvious pharmaceutical use for such substances. While many other cultures had traditionally included substances extracted from animals in their pharmacopoeias, Western medicine had relied almost exclusively on plant and mineral sources for its medications. The greatest success in the area of replacement therapy in the early twentieth century was the development of insulin therapy for the treatment of diabetes—up to that time a disease that almost certainly resulted in premature death. Another hormone, adrenaline, captured attention as an even earlier miracle drug, literally raising people from the dead, people whose hearts had stopped or who could not breathe because of asthma or severe allergic reaction—purposes for which adrenaline is still used.

This new knowledge could also be used to develop medicines that would counter the effects of the overproduction of one of the regulators or transmitters. Discovery stories often happened the other way around, however—beginning with known medicines and moving to biochemical and physiological understandings.

Others of today's familiar pharmaceuticals are also based in part on knowledge developed in the early twentieth century about the body's various regulators and transmitters. The first antihistamines used by allergy sufferers had appeared well before World War II (see Daniel Bovet, p. 68). All kinds of modern medicines to treat cardiovascular problems depend on information about the substances that control the autonomous nervous system. Knowledge of enzymes has often played a role behind the scenes in the development of new drugs. Recently, for example, drugs have been designed that inhibit production of an enzyme necessary for the AIDS virus to reproduce. Enzymes also serve as critical tools in today's biotechnology.

## LEONOR MICHAELIS (1875–1949) AND MAUD LEONORA MENTEN (1879–1960)

Enzymes are remarkable compounds that make possible the chemical reactions of life. These complex molecular machines work on a variety of other molecules, called substrates, to facilitate such changes as transfer of electrons (oxidation and reduction), molecular rearrangements, and water removal. Life goes on only because of the action of enzymes. The more that scientists understand the workings of enzymes, the better they understand the complex array of reactions that make up the web of life, in healthy and in diseased states. Seminal work published in 1912 by Leonor Michaelis and Maud Leonora Menten, a German man and a Canadian woman, cast light on the reasons why enzymes are so efficient.

The substances that came to be called enzymes were already being studied some seventy-five years before Michaelis and Menten's work. In the early nineteenth century (1835) Jöns Jakob Berzelius recognized that some substances in living bodies cause chemical reactions but do not themselves undergo change. He named these substances "catalysts" from the Greek word *katalyein* meaning to loosen or dissolve. Such substances were also called "ferments," which came in two varieties: "organized" ferments, which required the existence of a living body like yeast for their activity, and "unorganized" ferments, which are active even when removed from the organism. In the last quarter of the nineteenth century (1876) Willy Kühne coined the word *enzyme* (from the Greek word *enzymos* meaning "in yeast"), which Eduard Buchner applied to both sorts of ferments, since in 1897 he succeeded in extracting a substance from yeast ground with sand that continued to support fermentation. In 1894, in the course of his work on sugars, Emil Fischer noted the specificity of enzymes, that each enzyme works only with a specific substrate. He then made the often-quoted remark: "To make use of an image, I shall say that enzyme and glucoside must fit together like lock and key."

Leonor Michaelis received his medical degree at the University of Berlin in 1897, making clear his interest in research by doing an embryological study rather than a clinical study to complete his requirements. Immediately thereafter he wrote a brief textbook on embryology. The book was well received and went through several editions. The year after graduating he worked as an assistant in Paul Ehrlich's laboratory in Frankfurt. There he contributed to Ehrlich's program by studying the interaction between aniline dyes and the chemical constituents of living tissues. Michaelis also discovered a dye that specifically stained certain crucial components, later called mitochondria, of the cells that make up living tissue. In the same year he took up the study of the then-new subject of physical chemistry,

■ Maud Menten, pictured here in her youth. Courtesy Archives Service Center, University of Pittsburgh.

■ A portrait of Maud Menten. Courtesy Archives Service Center, University of Pittsburgh.

■ Leonor Michaelis in 1946. Courtesy Rockefeller University Archives.

which he was confident would have important applications in biological studies.

Michaelis followed Ehrlich's advice to proceed with clinical studies since finding support for fundamental scientific research was so difficult. After five years of clinical work at Berlin hospitals Michaelis was only able to obtain a courtesy appointment at the University of Berlin. He felt that as a Jew he would have little opportunity for advancement at the university, and so in 1905 he accepted a position as bacteriologist at one of Berlin's municipal hospitals. There he and a friend, chemist Peter Rona, set up a small research laboratory and conducted studies on the role of hydrogen-ion concentration in determining the properties of proteins and enzymes. These studies were similar to but independent of the work being done by the Danish chemist Søren Sørensen (whose definition of pH continues to be used today). Michaelis also studied how small molecules adhere to the surface of various proteins and enzymes. Michaelis then tried to understand the relation between the rate of formation of product to the concentrations of enzyme and its substrate—in other words, its efficiency in getting its work done.

In the early stages of this work he was joined by Maud Menten, one of about forty coworkers who were attracted to his modest laboratory during the period of its operation, from 1905 to 1921. (A subsequent postdoctoral worker was Albert Szent-Györgyi; see p. 30.) Menten had just received her M.D. in 1911 from the University of Toronto. Born in 1879 in Port Lambton, Ontario, Canada, she completed her bachelor's and master's degrees at the University of Toronto and then worked a year at the Rockefeller Institute in New York before returning to the university for her M.D.

Michaelis and Menten were able to express mathematically the relationship they were investigating, which demonstrated that each enzyme not only has its own substrate but also that at sufficient concentrations of substrate it has its own rate of causing that substrate to change chemically. One of the constants used in expressing this rate is now called the Michaelis-Menten constant. Much later, in

the 1930s, their work was highlighted by J. B. S. Haldane, a British physiologist, biochemist, and geneticist. Michaelis went on to study how enzyme activity is inhibited, a topic later of great importance to the development of pharmaceuticals (see Miguel Ondetti, p. 151, and Irving Sigal et al., p. 173).

When Menten returned from Berlin, she enrolled at the University of Chicago, where in 1916 she obtained a Ph.D. in biochemistry. Unable to find an academic position in her native Canada, in 1923 she joined the faculty of the medical school at the University of Pittsburgh while serving as a clinical pathologist at Children's Hospital in Pittsburgh. Despite the demands of these two positions Menten maintained an active research program, authoring or coauthoring more than seventy publications. Although her promotion from assistant to associate professor was timely, she was not made a full professor until 1949, when she was seventy years old and within one year of retirement.

Michaelis's career continued on a similarly slow track compared with his earlier achievement. In 1921 he joined the staff of a Berlin manufacturer of scientific equipment, and in 1922 he accepted an appointment as a visiting professor at the Aichi Prefectural Medical School in Japan, which later became part of the University of Nagoya. After three years there he came to the United States for a lecture tour and afterward stayed for three years at the Johns Hopkins School of Medicine. In 1929, at the age of fifty-four, he finally attained his objective of a permanent academic position, at the Rockefeller Institute in New York, where he again attracted a notable group of coworkers. There, with Ernst Friedheim, he made another important contribution to biochemistry, proving that the oxidation and reduction of organic substances often occurs in stepwise fashion, with the intermediate formation of free radicals. These structures contain one or more carbons with a single unpaired electron in addition to electron-pair bonds. At the time the notion of "free radical" was still very much the subject of skepticism.

# John Jacob Abel (1857–1938) and Jokichi Takamine (1854–1922)

In the late nineteenth century the search was on for the glandular products that in 1905 were named "hormones" by the English physiologist Ernest Henry Starling, a co-discoverer of secretin, which is secreted by the small intestine. The first hormone to be isolated in pure form was epinephrine (popularly known as adrenaline). The competition to isolate epinephrine engaged, among others, John Jacob Abel and Jokichi Takamine, scientists from quite different backgrounds who were both working in the United States and who both counted accomplishments in broad areas of science and society.

John Jacob Abel was born on a farm near Cleveland, Ohio, to parents of German origin. His mother succumbed to puerperal (childbed) fever after the birth of her eighth child, when John Abel, the eldest, was just fifteen. Somehow the family managed, with older children taking care of younger ones, so that Abel could attend Cleveland public schools and three years at the University of Michigan. Then he had to drop out to earn money. In a day when credentials for teaching school were minimal, he served as principal of the high school and then superintendent of schools in La Porte, Indiana. In those years he decided to become a physician and met an English teacher who was to become his wife, Mary Hinman. On Abel's return to the university for his bachelor's degree he was permitted to take some of his courses in the university's medical school. Always more interested in research than in clinical practice, he set out in 1883 for graduate study at the recently founded Johns Hopkins University, which did not yet have a medical school.

As was common for Americans until World War I, Abel pursued further study in Europe. He first went to Leipzig, then to Strassburg (where the Abels' first child died of polio), then to Vienna, and finally to Bern. During his protracted studies he earned an M.D. (in 1888) and worked with some of the leaders in the new field of pharmacology. This field, which brought physiology and chemistry to bear on the effects of drugs on living systems, became essential to the pharmaceutical research process in testing for the safety and effectiveness of potential new therapies.

Finances were always a problem for the little family. Mary helped their situation by winning an essay contest on the set topic of practical sanitary and economic cooking for the moderate-income family. Mary's essay brought her to the attention of Ellen Swallow Richards, the first female instructor at the Massachusetts Institute of Technology and a pioneer in home economics and environmental studies. After Mary returned to the United States, she ran Richards's New England Kitchen in Boston for six months and shipped funds over to her husband, who was still in Europe. Later the two women became pillars of the newly founded American Home Economics Association.

Finally, in 1891, Abel was appointed to the medical school at the University of Michigan to teach *materia medica* (medical materials) and therapeutics and perhaps some physiological chemistry. Although *materia medica* suggests the teaching by rote of knowledge about drugs then in use and their supposed effects on patients, Abel transformed the course into an introduction to pharmacology, complete with experimentation on animals. He had barely gotten resettled in Michigan, when, in 1893, he was asked to join the faculty of the recently opened Johns Hopkins Medical School as professor of pharmacology—the first professorship of pharmacology in the United States.

Eventually, Abel came to be regarded as the "father of U.S. pharmacology" because of the large number of leading pharmacologists he educated at Hopkins and the role he took in founding a professional society and related journals. He was valued by his postgraduate students because of the great interest he took in his discipline, evident in conversation at the communal lunch table, and by the example he set in the laboratory as the devoted researcher, persevering even after losing sight in one eye in a laboratory explosion.

Abel's most celebrated research was really more biochemical than pharmacological; that is, in this research, only agents that occurred naturally in the body were involved. He joined in the late-nineteenth-century competition to isolate hormones, first trying to find the active principle in thyroid gland extract. He abandoned the chase, however, when in 1895 Eugen Baumann isolated an iodine-containing organic compound from that gland (although, as it turned out, he had not reached the objective either). In the same year George Oliver and Edward Schäfer of University College, London, prepared an extract from the adrenal medulla (the adrenal gland's inner part) that produced an immediate and

■ John Abel, holding a syringe and test tube. Courtesy Alan Mason Chesney Medical Archives of the Johns Hopkins Medical Institutions.

significant increase in the blood pressure of experimental animals injected with the extract. In 1897, by using various separation techniques, Abel managed to obtain a crystalline product from sheep adrenal glands, which he named "epinephrine," that was remarkably effective in raising blood pressure. He had not, however, isolated the pure substance but probably a slightly impure derivative of it.

At this point Jokichi Takamine enters the story. Takamine was by 1901 living and working in New York City and Clifton, New Jersey, where he retained the services of a Japanese chemist, Keizo Wooyenaka, to assist him in obtaining the pure substance secreted by the adrenal medulla. They succeeded, and Takamine obtained patents for the substance, which he named "adrenaline," as well as for his process of obtaining it. (Both Takamine's and Abel's names for the hormone refer to the gland "on top"—(*ad* in Latin, *epi* in Greek—and to the "kidneys"—*renes* in Latin, *nephros* in Greek.) Using combustion analysis—burning the substance and weighing the products—Takamine attempted but failed to determine the correct chemical formula for adrenaline, as had Abel; the correct formula ($C_9H_{13}NO_3$) was actually determined by Thomas Aldrich. Aldrich was a chemist working at Parke, Davis and Company in Detroit, a company with which Takamine had had a long business relationship (and now a division of Pfizer Inc.). Takamine sold the right to produce adrenaline to Parke, Davis, which trademarked it as Adrenalin. It rapidly came into clinical use in relieving respiratory distress, as in asthma attacks or allergic reactions, and to start up the hearts of patients who have had cardiac arrest.

In a 1911 landmark court case the H. K. Mulford Company of Philadelphia (acquired by Merck and Company in 1929) attacked the adrenaline patent on the grounds that the hormone exists in nature and that Takamine's work had been anticipated by Abel and others. It was even alleged that, in a visit to Abel's laboratory at Johns Hopkins, Takamine had obtained information critical to his success without acknowledging its source. After listening to days of technical testimony, Learned Hand, then federal district judge in New York and already famous for his judicial wisdom, remarked: "I cannot stop without calling attention to the extraordinary condition of the law which makes it pos-sible for a man without a knowledge of even the rudiments of chemistry to pass upon such questions as these." Judge Hand ruled in favor of Takamine.

Because the cost of extracting epinephrine from glandular material is exorbitant, the pharmaceutical industry has long searched for substitutes. Roughly contemporary with Abel's and Takamine's work was the isolation of ephedrine from the ancient Chinese herbal drug *ma huang* by Nagajosi Nagai in 1897 and independently by Ko Kuei Chen in 1923. Chen had been trained by Abel at Johns Hopkins, and in 1929 he was to become George Clowes's successor as research director at Eli Lilly and Company (see Clowes, p. 57). Ephedrine and one of its forms, pseudoephedrine, are commonly found today in asthma medicines and decongestants. In 1932 Gordon Alles, who worked at Smith, Kline and French, synthesized another adrenaline substitute, amphetamine. The company marketed it as Benzedrine, without understanding the addictiveness of their product. (Methamphetamine, a related chemical, is the notorious "speed.")

As this little excursion into substitutes for adrenaline demonstrates, Takamine was one of several Asian scientists who began to make their mark on Western science at the end of the nineteenth century. He was born in Kanazawa, in northern Japan, to a samurai family whose members traditionally served as physicians rather than warriors. Takamine's parents wanted him to be part of the new era of intellectual and commercial exchange between Japan and the West that blossomed in the second half of the nineteenth century. As a boy, Takamine was sent to live with a Dutch family in Nagasaki in order to learn English, which he spoke with a Dutch accent for the rest of his life. He attended the Imperial Engineering College in Tokyo and graduated as a chemical engineer in 1878. He was then sent by the Japanese government to continue his education at the University of Glasgow and Andersonian College, also in Glasgow, where he studied fermentation processes and picked up a slight Scottish burr.

When Takamine returned to Japan, he joined the Imperial Department of Agriculture and Commerce, where he soon became director of the Government Chemical Laboratory. His specialty in the laboratory was investigating the

■ Jokichi Takamine. Courtesy Bayer Corporation.

problems encountered by brewers who produced sake, an alcoholic beverage made from rice and usually served hot. In 1884 Takamine led the Japanese delegation to the International Cotton Centennial Exposition held in New Orleans, where he met the woman who would become his wife, Caroline Field Hitch. The couple returned to Japan, where Takamine put his chemical engineering training to work in establishing the first plant in Japan to produce superphosphate fertilizer.

In the late 1880s he pursued his own research to improve the process that the Japanese had long used to convert the starch in rice to sugars, which could then undergo fermentation to alcohol by the action of yeast. In the West, brewers traditionally converted grains to sugars before fermentation by adding malt (sprouted barley or other grain), which produces groups of enzymes called diastase or amylase that liquefy and split up the starch. In Japan a far-more-active diastase was derived from moldy rice. Takamine carefully selected the diastase-producing mold, *Aspergillus oryzae*, and for rice, he substituted wheat bran, the usually discarded by-product of flour production. He predicted that his cheaper and more effective enzymes would revolutionize the brewing industry worldwide.

In 1890 his American father-in-law financed the return of Takamine and his young family to the United States so that he could try his luck as an entrepreneur. He established the Takamine Ferment Company in Peoria, Illinois, to produce enzymes for a local distiller. In 1894 he received a U. S. patent for Taka-Diastase, the same year that the distillery with which he had his business relationship burned to the ground—probably the work of arsonists acting on anti-foreign rhetoric encouraged by rival malt producers and distillers. Takamine then moved to Chicago and sold the patent rights for Taka-Diastase to Parke, Davis, which marketed the enzyme as a way to prevent indigestion supposedly caused by a deficiency in ptyalin, the starch-digesting enzyme in saliva. In 1897 Takamine moved his operations to New York and New Jersey, founding the International Takamine Ferment Company and Takamine Laboratory, where the adrenaline research took place. (Ultimately, after

the death of his youngest son, Takamine's American businesses were purchased by Miles Laboratory, now Bayer.)

With his fame and fortune secure because of sales of adrenaline, Takamine worked to bring Japan's industries into the twentieth century, an effort that included such ventures as the formation of the Sankyo Pharmaceutical Company and a tour of Japan with his friend Leo Baekeland to introduce plastics and their production techniques to the Japanese.

Takamine also strove to increase American appreciation of Japanese culture. Symbolic of this endeavor was his enthusiastic support of First Lady Helen Herron Taft's plan to beautify the Tidal Basin area of Washington, D.C. In 1911 Takamine arranged for the gift of three thousand cherry trees from the mayor of Tokyo for this project, which Washingtonians and tourists enjoy to this day. At the same time hundreds of cherry trees were sent to New York City and Detroit (home of Parke, Davis).

Meanwhile, Abel, in the years immediately following his work on epinephrine, became deeply interested in the protein constituents of the blood. During this research he devised an artificial kidney—an early dialysis machine—so that such substances could be removed from the blood of a living animal for study and the blood itself could be returned to the animal's circulation system. He predicted that such machines would some day be used to help human patients suffering from kidney disease.

Abel eventually returned to his early interest in hormones and in 1924 was called upon by his friend Arthur A. Noyes at the California Institute of Technology to attempt to isolate a purer insulin than Frederick Banting and James Macleod and their associates had been able to achieve in Toronto (see Banting et al., p. 52). Noyes and Abel hoped thereby to understand the nature of insulin better. Abel's claims that he had crystallized the pure hormone and that his results suggested that insulin is a protein first met with great skepticism. This attitude was changed only by subsequent evidence that many more body substances, including the enzymes, are actually proteins than anyone had previously thought.

# FREDERICK GRANT BANTING (1891–1941), CHARLES HERBERT BEST (1899–1978), JAMES BERTRAM COLLIP (1892–1965), AND JOHN JAMES RICKARD MACLEOD (1876–1935)

At the University of Toronto, at the end of the 1920–21 school year, researchers began a series of experiments that would ultimately lead to the isolation and commercial production of the pancreatic hormone insulin and the successful treatment of diabetes.

First recorded by physicians in ancient Egypt, diabetes mellitus, a disease that exists in mild to severe forms, is characterized by frequent urination, extreme thirst and hunger, and in the worst cases, death by starvation. Pre-twentieth-century treatments for diabetes varied, from the prescription of opium to the prescription of a sugar-rich diet. Physicians eventually found that opium did not help in treating the disease, and sugar-rich diets actually speeded the decline and demise of diabetics. During the Prussian siege of Paris in the 1870s a French doctor noticed that the symptoms of some of his diabetic patients actually disappeared under the extreme rationing forced on inhabitants of the city. Gradually, doctors came to realize that high-calorie diets were devastating to the health of the diabetic, and new regimens of low-calorie (sometimes as low as 500 calories per day), no-carbohydrate diets were prescribed. By 1921 this strict diet regulation was the only effective treatment of diabetes. Of course, this method had its consequences, as slow starvation, like diabetes, drained patients of their strength and energy, leaving them semi-invalids. The diet treatment also required an inordinate amount of willpower on the part of the patient, very few of whom were able to maintain low-calorie diets long term.

The connection between pancreatic secretions and diabetes was shown in 1889 by the German physiologists Oskar Minkowski and Joseph von Mering, who were working at the University of Strasbourg. While investigating the effect of pancreatic secretions on the metabolism of fat, Minkowski and von Mering performed a complete pancreatectomy on a laboratory dog, only to discover that the animal developed a disease indistinguishable from diabetes. Twenty years earlier a German medical student, Paul Langerhans, had reported that during the course of his dissertation research he had discovered two systems of cells in the pancreas. One set of cells, called the acini, produced the pancreatic digestive secretions. Langerhans admitted that the function of the other system of cells, which appeared to be tiny clusters, or islands, of cells floating among the acini, was unknown to him. A few years later these cells were named "islets of Langerhans" in acknowledgment of their appearance and their discoverer.

In 1901 Eugene Opie, an American pathologist at Johns Hopkins University, made the association between the degeneration of the cells of the islets of Langerhans and the onset of diabetes. Through the experimental efforts of these and many other researchers, the stage was set for the discovery of insulin (the hormonal antidiabetic secretion of the islets of Langerhans) in the first decades of the twentieth century.

In 1920 Frederick Banting was a surgeon in a floundering practice in London, Ontario, Canada. The youngest son of Methodist farmers from Alliston, Ontario, Banting almost entered the Methodist ministry but decided at the last moment that his calling lay in medicine. World War I shortened his five-year medical course: his class did its entire fifth year during the summer of 1916 and, upon receiving their hasty degrees, went off to war. Banting served as a battalion medical officer in the Canadian Army Medical Corps; he returned to Toronto in 1919 after having been wounded in the arm by shrapnel. He trained as a surgeon at the Hospital of Sick Children in Toronto, then decided to open a small practice as a surgeon in London, Ontario. Unfortunately, his earnings from his practice were meager, forcing him to take a position as a demonstrator in the local medical school. It was in this capacity that Banting was preparing a lecture about the function of the pancreas on 30 October 1920. He stopped at the medical school library, where he picked up the latest issue of *Surgery, Gynecology and Obstetrics*, and read an article titled "The Relation of the Islets of Langerhans to Diabetes, With Special Reference to Cases of Pancreatic Lithiasis."

While thinking about pancreatic secretions after reading the article, Banting jotted a preliminary experiment in his lab notebook. On 7 November, following the advice of a colleague, Banting brought his idea to the attention of John James Rickard Macleod, a Scottish physiologist and expert in carbohydrate metabolism at his alma mater, the University of Toronto. Macleod, the son of a minister, received his medical training at the University of Aberdeen and his biochemical training at the University of Leipzig. In 1903

■ Frederick Banting and Charles Best on the roof of the University of Toronto's Medical Building in 1922. Insulin tests were performed on Marjorie, the dog in the photograph. From the F. G. Banting Papers. Courtesy Thomas Fisher Rare Book Library, University of Toronto.

Macleod emigrated to the United States to take a position as professor of physiology at Western Reserve University (now Case Western Reserve University) in Cleveland. After fifteen years at Western, Macleod accepted a professorship at the University of Toronto, where he conducted research on respiration.

Earlier in his career Macleod had published a series of papers on glycosuria, or the presence of sugar in the urine (a common indication of diabetes). As a scientist familiar with the literature on the subject, he was unimpressed with Banting's range of knowledge about diabetes and the pancreas and skeptical about the soundness of Banting's idea.

■ John Macleod in 1923. From the C. H. Best Papers. Courtesy Thomas Fisher Rare Book Library, University of Toronto.

However, Macleod decided to give him lab space, an assistant, and some laboratory dogs for two months at the end of the academic year.

Banting and his assistant, Charles Best, began their experiments in May 1921. Best, the American son of Canadian parents, had just finished his bachelor's degree in physiology and biochemistry at the University of Toronto and had been hired as a research assistant to Macleod, his former teacher. Macleod assigned him to Banting, and the twenty-nine-year-old surgeon and the twenty-two-year-old assistant began their work together.

A combination of timing and good luck enabled the Toronto researchers to be the first to announce the discovery of insulin. Scientists in Germany and Hungary had come very close to finding pure insulin, but lack of funding and the devastation of World War I halted their progress. Following in the footsteps of earlier researchers, Banting and Best began to study diabetes through an experimental combination of duct ligation, which involved tying off the pancreatic duct to the small intestine, and pancreatectomies, or the complete surgical removal of the pancreas. Duct ligation served to atrophy the pancreatic tissues that produced the external, digestive secretions, leaving behind the cells of the islets of Langerhans. Duct-ligated dogs, it was discovered, did not develop diabetes. Pancreatectomy was the method of inducing diabetes: when all pancreatic tissue was removed, the experimental dogs immediately showed signs of glycosuria.

Banting's idea of 30 October involved ligation of the pancreatic ducts of a dog and the extraction and isolation of whatever secretions were produced after the atrophy of the acini cells. He and Best began this experiment, only to find that it was difficult to keep duct-ligated, de-pancreatized dogs alive long enough to carry out any tests. After a summer of many setbacks and failures, however, the team reported in the fall that they were keeping a severely diabetic dog (affectionately named Marjorie) alive with injections of an extract made from duct-ligated pancreas and prepared, following Macleod's instructions, in saline. Amazingly, this extract—the internal secretion of the pancreas—dramatically lowered the blood sugar levels of diabetic experimental dogs.

On 30 December 1921 Macleod, Banting, and Best presented their findings at the conference of the American Physiological Society, at Yale University. Banting, out of nervousness and inexperience, did a poor job delivering the paper, and the audience was highly critical of the findings presented. Macleod, as the chair of the session, joined the discussion in an attempt to rescue Banting from the scathing commentary. As one audience member recalled, years later: "Banting spoke haltingly, Macleod beautifully." After this fiasco Banting became convinced that Macleod had stepped in to steal the credit from him and Best, and relations between the two began to deteriorate.

At the end of 1921 Macleod invited James Collip, a biochemist in the department of physiology at the University of Toronto, to help Banting and Best with purifying their extract. Collip, another University of Toronto graduate, was on sabbatical from the University of Alberta and, supported by a Rockefeller Foundation Traveling Fellowship, had returned to his alma mater. As the experimental pace quickened, Banting and Best needed large amounts of their extract, and Collip, as a biochemist, set to work purifying the extract for clinical testing in humans.

The first clinical tests on a human patient were conducted on a severely diabetic fourteen-year-old boy. Although the injections of the extract failed to have resoundingly beneficial effects, the Toronto team continued to experiment. A short while later Collip made a breakthrough in purifying the extract, using alcohol in slightly over 90-percent concentration to precipitate out the active ingredient (insulin). At the same time, though, personal tension was mounting between the four scientists, as Banting became increasingly bitter toward Macleod and pitted himself and Best against Collip in the race to purify the extract. At the end of January, Collip came to Banting and Best's lab and informed the two that, although he had discovered a method to produce pure extract, he would share it only with Macleod. It was only Best's quick restraint that stopped Banting from attacking Collip. Fortunately for the future of insulin an uneasy agreement made a few days later allowed them to continue to work together. On 3 May 1922 Macleod, representing the group, announced to the international medical community at a meeting of the Association of American Physicians that they had discovered "insulin"—the antidiabetic agent.

Banting and Macleod received the 1923 Nobel Prize in physiology or medicine for the discovery of insulin. That the Nobel committee chose only Banting and Macleod for the award caused more animosity—Banting, outraged that Macleod was chosen to share the prize with him, immediately announced that he would split his winnings with Best. Macleod, perhaps in reaction to Banting's gesture, announced that he, too, would be splitting his award, with

■ James Collip at the lab bench in 1915. From the J. B. Collip Papers. Courtesy Thomas Fisher Rare Book Library, University of Toronto.

■ A letter to Banting from insulin recipient Myra Blaustein, dated 21 December 1922. From the F. G. Banting Papers. Courtesy Thomas Fisher Rare Book Library, University of Toronto.

Collip. By the end of 1923 insulin had been in commercial production for a year at the Eli Lilly and Company laboratories in Indianapolis. Diabetics who received insulin shots recovered from comas, resumed eating carbohydrates (in moderation), and realized they had been given a new lease on life.

In 1955 Frederick Sanger, a British biochemist, determined the exact ordering of the amino acids in insulin. This discovery was followed by Dorothy Crowfoot Hodgkin's determination of the three-dimensional structure of insulin in 1969. An extremely complex protein, insulin was extracted from tons and tons of animal pancreases until 1996, when the first analog of human insulin was developed using DNA recombination technology. Currently, both animal insulin and human analog insulin are used to treat diabetics.

Insulin by no means "cures" diabetes. Diabetics must combine their insulin injections with a carefully regulated diet and, even so, often still suffer many of the complications associated with diabetes, including impaired vision, kidney disease, and circulatory problems, to name a few. But increasing knowledge about the different forms of the disease and their causes—whether dietary or genetic—promises new avenues of treatment, prevention, and perhaps cure.

# George Henry Alexander Clowes (1877–1958)

The first public report of the pancreatic extract later known as insulin came on 30 December 1921 at a conference of the American Physiological Society held at Yale University. As Banting, Best, and Macleod reported their experimental findings to the North American medical community, George H. A. Clowes, the research director of Eli Lilly and Company of Indianapolis, sat in the audience, captivated by what he was hearing. Clowes had been tipped off about the insulin research a month earlier by a colleague of Macleod's and had decided to attend the December meeting where the Toronto team would be presenting their findings.

To the great fortune of diabetics, Clowes was convinced that the Toronto team had discovered an immensely important drug. Immediately after the conference Clowes telephoned Macleod and proposed collaboration between the Toronto team and Eli Lilly in preparing the extract commercially. Banting, Macleod, and Clowes discussed the idea, but the researchers felt further experimental work was needed before the extract could be put on the market.

Clowes, an Englishman with a Ph.D. in chemistry from the University of Göttingen, did his postdoctoral work at the Sorbonne (he firmly believed that all English-speaking chemists should be fluent in both French and German). Some years later the opportunity to join a cancer research program at a New York State research institute induced Clowes to emigrate to the United States. During World War I, he joined the U.S. Army and studied the physiological effects of mustard gas. In 1918 this research brought Clowes to the Marine Biological Laboratories in Woods Hole, Massachusetts, where he observed the effects of the gas on marine organisms. This was the beginning of a longstanding relationship with the Marine Biological Institute; Clowes returned to Woods Hole every summer for the next forty years to continue his work with marine organisms and his research on cancer. Clowes joined Eli Lilly and Company in 1919 as a special research chemist. A year later he was appointed research director of the company. Throughout his forty-year career at Lilly, Clowes was committed to strengthening ties between industry and the larger scientific community—a determination that brought him to Toronto.

On 30 May 1922, Clowes visited the Toronto team again. The team had recently announced the clinical success of insulin in human diabetics, and impressed by the plans laid out by Clowes for the commercial production of insulin, it agreed to give Lilly an exclusive license for one year. The collaboration between Lilly and the University of Toronto team was formally called an "indenture" and was meant to establish a close, but not necessarily exclusive, relationship between the two entities. Plans were drawn up specifying that for one year the Toronto researchers and the Lilly researchers would share all experimental findings and any improvements made in the production process. During this experimental year large-scale clinical testing would be conducted in Canada and the United States, and Lilly would supply the insulin free of cost during the initial stages and later at cost. When the year was over, Lilly would have a license to manufacture and market insulin on the same terms that Toronto would license other manufacturers. Insulin licensees, it was agreed, would pay royalties on all insulin sold to the University of Toronto. The terms of the contract specified that Lilly would have sole rights to manufacture and market insulin in the Americas; the Toronto team decided to offer insulin patent rights for Britain, its empire, and potentially the rest of Europe to the British Medical Research Council.

This agreement made, Clowes assigned two research chemists to the task of developing a method for mass-production that did not diminish insulin's potency. The problem of weakened potency had plagued the Toronto team as they attempted to produce larger and larger quantities of the hormone. Lilly's husband-and-wife team, George B. and Eda Bachman Walden, both chemists, solved this problem by adding a special isoelectric precipitation step that ensured a potent product of standardized purity.

Iletin, Lilly's brand of insulin, was introduced to the commercial market in October 1923, after a year of extensive clinical testing. Clowes's sound investment not only brought remarkable sales in the first year, but more important, also offered hope to diabetics around the world.

■ George Clowes, with the world in hand.
Courtesy Eli Lilly and Company Archives.

## SIR HENRY HALLETT DALE (1875–1968) AND OTTO LOEWI (1873–1961)

After a lifetime of friendship and scientific camaraderie Sir Henry Hallett Dale and Otto Loewi were jointly awarded the Nobel Prize in physiology or medicine in 1936 for their revolutionary contributions to the understanding of the *chemical* transmission of nerve impulses. That theory was first put forward by T. R. Elliott in 1904, but until Dale and Loewi's work the dominant theory emphasized the electrical nature of communication between nerves.

Dale and Loewi also made contributions to understanding histamines. Much of their work concerned the autonomic nervous system, which controls the body's organs unconsciously and is made up of the sympathetic and parasympathetic nerves (the latter generally acting to oppose the sympathetic nerves). Since these are among the systems that keep us alive day in and day out, the implications of being able to affect them predictably through pharmaceutical intervention are evident. Although Dale and Loewi's direct contributions to therapy were relatively minor, such possibilities were exploited by their successors. (See, for example, Daniel Bovet, p. 68.)

Henry Hallett Dale was born in London in 1875, one of seven children. His father, a businessman and prominent Methodist, presumed that his son would follow in his footsteps until Henry's teachers pointed out his interest in and aptitude for the sciences. In 1894 Dale enrolled in Trinity College, Cambridge, funding his education through scholarships and other financial aid. He then pursued his medical studies at St. Bartholomew's Hospital in London and in 1909 earned his medical degree.

Dale and Loewi first met just after the turn of the century, in the research laboratories of Ernest Henry Starling, a well-respected British physiologist, at London's University College. Dale, who spent two years there on a research studentship, had been given a research room in Starling's department and the opportunity to work on the effect of prolonged stimulation of secretin on the pancreas. Secretin, a digestive hormone secreted by the small intestine, had just been discovered by Starling and Sir William Maddock Bayliss. Although Loewi came to England from Germany to work with Starling only briefly, he left having made a lifelong friend in Dale. The two scientists met again in 1903, when Dale arrived in Germany to work with Paul Ehrlich (see p. 17), then the director of the Institute of Experimental Therapy. Dale went directly to see his friend, who traveled with Dale to Ehrlich's institute in Frankfurt.

Dale spent only four months in Germany (the work he was asked to do at the institute held little interest for him), but he was fortunate enough to have witnessed the beginnings of Ehrlich's now-famous work on chemotherapy. Combined with his previous experience with Starling, his exposure to Ehrlich's style of work provided Dale with a good background in physiological research, his chosen profession.

When Dale returned to England in 1904, he accepted a research position at the decade-old Wellcome Physiological Research Laboratories, against the advice of his many scientific counselors. Although it took him away from the standard academic track followed by many of his colleagues, Dale saw the position as an opportunity to conduct his own research.

On arriving at Wellcome Laboratories, Dale found himself without any particular research project, and Henry S. Wellcome—founder of the laboratories—suggested that he investigate the pharmacology of ergot, because one of Wellcome's competitors, Parke, Davis and Company, was attempting to standardize various ergot derivatives. Ergot is a disease of cereal grains caused by the fungus *Claviceps purpurea*, and the diseased seeds themselves are also called ergot. Products derived from plants infected with the disease had long been used for a variety of medical purposes, including the treatment of migraines and the control of postpartum hemorrhage.

Using ergotoxine, a mixture of substances produced by ergot and extracted by George Barger (see below), Dale was testing for "pressor action"—the ability to increase blood pressure—on a cat. His experiment was interrupted by a delivery of dried adrenal glands from the Burroughs-Wellcome factory, which he was supposed to test for adrenaline content—testing substances for their activity in living systems being one of the more routine functions of the lab. As the cat was already there, he decided to test a sample on it even though it had already been dosed with ergotoxine. Rather than raising the cat's blood pressure, as adrenal-gland material should, it lowered it. Dale at first assumed that the adrenal-gland sample was defective until this "lucky accident," as he later called it, was repeated a week later. In the

■ Henry Dale (center), with Richard Birks and F. C. McIntosh, examining an electron microscope at McGill Medical School, in Montreal, 1959. The microscope was funded by the Wellcome Trust, which Dale headed at the time. Courtesy GlaxoWellcome Inc. Heritage Center.

resulting paper published in 1906, Dale announced that ergotoxine could reverse adrenaline's pressor action and its stimulation of sympathetic nerves to the heart and elsewhere, thereby introducing the first adrenergic-blocking agent, a precursor of several of today's blood pressure medications. By 1910 a little structural chemistry had entered into the discussion when Dale and Barger were able to show that other amines with chemical structures similar to adrenaline produced effects similar to ergotoxine's.

Dale carried out still other work on ergot with Barger, Arthur Ewins, and P. P. Laidlaw, much of it done at the laboratory to which he moved in 1914 and which later became part of the National Institute of Medical Research (NIMR). Their research revealed that ergot contains a number of pharmacologically active substances, including what are now called histamines and acetylcholine. They carefully investigated the physiological effects of these ergot-derived substances.

The histamines acted on smooth muscles, capillaries, gland cells, and the central nervous system. Acetylcholine had multiple effects. It slows the heart by stimulating parasympathetic nerves that oppose the sympathetic nerves that prepare the body for stress by increasing the heart rate and the flow of blood to the muscles. However, when acetylcholine's effects were blocked by atropine—an alkaloid derived from the deadly nightshade plant, with physiological effects similar to adrenaline—it stimulated the sympathetic nerves of the autonomic nervous system. Only later did they recognize that acetylcholine also has effects on the junction of nerves to voluntary muscle.

Despite this extensive investigation proof was long in coming for the suspicion held by Dale and others that these substances occur naturally in animal bodies. Dale's supposition, however, that histamines were at work in human tissues enabled him to understand and suggest successful means to treat "wound shock" during World War I. Wound shock is the alarmingly sudden and often lethal failure of the circulatory systems of seriously injured people, whose wounds have neither bled excessively nor become infected. Dale, working with Laidlaw and then with A. N. Richards, a University of Pennsylvania researcher he had asked to join them, explained that the catastrophic decline was caused by an actual deficiency in the amount of circulating blood; the injured tissues had released something (later identified as histamines) that caused capillaries to dilate and to allow blood to collect while plasma escaped through the walls of the blood vessels. The most immediately effective remedy was, and is, to give blood or plasma transfusions.

Dale's early research on acetylcholine provided some support for the hypothesis of chemical transmission of nervous impulses, but the most convincing evidence was produced by Dale's friend Otto Loewi.

Loewi was born in Frankfurt in 1873. Although he displayed an interest in the humanities (particularly art history) during his early education, his parents encouraged him to pursue medicine. He attended the Universities of Munich and Strassburg, emerging in 1896 from Strassburg with his medical degree. Loewi then spent some time in clinical medicine, a subject he soon discarded in favor of pharmacological research. He joined the laboratory of pharmacologist Hans Horst Meyer in 1898, then left eleven years later to become chair of pharmacology at the University of Graz.

In his wide-ranging physiological and pharmacological research, Loewi had at various times examined the physiology of the frog's heart, but it was an experiment in 1921 that verified the chemical transmission of nerve effects. Inspired by a dream, Loewi arose in the middle of the night to conduct the experiment. It entailed isolating two frog hearts, one with nerves and one without; attaching both to small containers for holding nutrient solutions; stimulating the vagus nerve of the first heart (which slowed the heart rate); and transferring the nutrient solution from the first heart to the second heart. The result was that both hearts slowed. Then the accelerator nerve in the first heart (which sped up the heart rate) was stimulated, and the nutrient solution transferred again. The result was that both hearts sped up. These results indicated that it was not the nerves themselves but rather chemical substances that produced the reactions.

Despite this great breakthrough, five years passed before Loewi, along with his collaborator, Ernst Navratil, positively identified the substance produced by the vagus nerve as acetylcholine. During this time the scientific community did not accept his frog-heart experiment, partly because electrophysiologists were reluctant to accept a theory of

■ Otto Loewi. Courtesy New York University Archives.

chemical rather than electrical transmission but mostly because the experiment was difficult to reproduce, even by Loewi himself. However, after he successfully demonstrated the experiment at the Twelfth International Congress of Physiology in Stockholm in 1926, this criticism dissipated. By 1936 Loewi had conclusively determined that the accelerator substance in the frog-heart experiment was adrenaline.

Dale, meanwhile, was following Loewi's careful work with interest while successfully cooperating with H. W. Dudley in the search for acetylcholine in normal animal tissue. They found acetylcholine in mammalian spleen in 1925 and the histamines in fresh liver and lungs in 1927. In the early 1930s Dale and his colleagues were able to measure the amount of acetylcholine released after nerve stimulation in various sites in both the parasympathetic and the sympathetic nervous systems. For this purpose they experimented on leech muscle that had been treated with an alkaloid extracted from the Calabar bean, physostigmine, used in the treatment of glaucoma (see Percy Julian, p. 106). Since some acetylcholine is normally destroyed by the body, they used physostigmine to prevent this destruction, thereby allowing high levels of acetylcholine to build up, which could then be measured.

By this period in his career Dale had become the first director of the NIMR. In addition to his continued research on acetylcholine he made enduring scientific contributions by leading the fight to standardize dosages of various medications, including insulin (see Frederick Grant Banting et al., p. 52). Through his extensive efforts, in 1925 the Geneva meeting of the International Conference on Biological Standards defined a unit of insulin as a precise weight of a dry, stable sample (eight units per milligram).

At his retirement from the NIMR in 1942 Dale was already serving as the president of the Royal Society, one of several leadership roles in science that he assumed over his lifetime. Another was as trustee and chair (from 1938 to 1960) of the Wellcome Trust, set up in 1936 by Henry Wellcome's will to sponsor medically related research independent of the Burroughs-Wellcome Company.

In 1938, just two years after winning the Nobel Prize, Loewi, a Jew, was imprisoned by the Nazis. He was released and permitted to travel to London, where Dale had arranged a cordial welcome for him. Meanwhile Loewi's wife was detained until the couple's assets, including his Nobel Prize money, had been turned over to the Nazis. Loewi worked at various European research institutes until 1940, when he secured a position at New York University's College of Medicine, where he remained until his death.

# 5. Sulfa Drugs

Although Ehrlich's salvarsan (Figure; 1912) had proved effective against syphilis, it was ineffective against other harmful bacteria. Efforts to find chemotherapeutic treatment for a number of tropical diseases caused by protozoa, like malaria, met with success, but common bacterial infections that were more prevalent in Europe and the United States still ran rampant.

In the 1920s and 1930s staphylococcal and streptococcal infections loomed large as killers along with pneumococcal and tubercular infections. In this environment minor scratches and scrapes could prove deadly, and pneumonia and tuberculosis killed even young adults, generally the most vigorous segment of a population. Introduced in 1935, the sulfa drugs, or sulfonamides, all of which are related to the compound sulfanilamide, provided the first successful therapies for many bacterial diseases.

As such, they proved to be the forerunners of antibiotics, showing that bacterial diseases are indeed vulnerable to substances not natural to the human body and proving the necessity of testing the safety and effectiveness of such drugs. Although antibiotics have by now nearly monopolized the treatment of bacterial diseases, certain sulfa drugs are still commonly used to knock out infections quickly. For example, middle-ear infections in children are often treated with Pediazole, a combination of an antibiotic and a sulfa drug; and Bactrim, which contains a sulfa drug, is prescribed for urinary tract infections in patients of all ages.

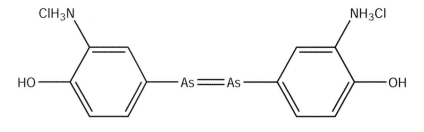

**Figure.** Salvarsan.

# GERHARD DOMAGK (1895–1964)

In 1927 Gerhard Domagk, a thirty-two-year-old M.D., joined the research staff at the pharmaceuticals division of IG Farbenindustrie, a recently organized industrial conglomerate made up of German companies, including the former Bayer Company, that specialized in dyes (*Farben*) and other fine chemicals. Heinrich Hörlein, the director of several laboratories in the company's Elberfeld locations, hired Domagk to establish a special pharmacology laboratory and to collaborate with two chemists, Fritz Mietzsch and Josef Klarer. They worked together on a research program to test compounds related to synthetic dyes for their effectiveness against disease—a program based on Paul Ehrlich's example (see p. 17) but expanded and systematized to the extent that it could be accurately termed "industrial research." Mietzsch had already accomplished his synthesis of Atabrine, a successful substitute for quinine, the natural antimalarial extracted from the bark of cinchona trees. Since these trees grew almost exclusively in the Dutch East Indies, natural quinine was subject to monopolistic pricing and shortages in wartime. Finding effective treatments for bacterial diseases was very much on the new team's research agenda.

Born the son of a teacher in Lagow in Germany's province of Brandenburg, Domagk decided early in life to become a physician. His medical studies at the University of Kiel were interrupted by his service as a grenadier and medical corpsman in World War I. He completed his medical degree in 1921 and then began an academic career, pursuing research in pathology. As his experiments on the functioning of the immune system in mice demonstrated, he had adopted a dynamic approach to pathology, incorporating physiology and chemistry into his work. Domagk was recruited from the University of Münster by IG Farben, where he spent the rest of his career. (After World War II, IG Farben was broken up, and the division in which Domagk worked became once again the Bayer Company.) There, he not only felt at greater liberty to pursue his research but also had far better resources at his command.

In Domagk's view a drug's role was to interact with the immune system, either to strengthen it or so weaken the agent of infection that the immune system could easily conquer the invader. He therefore placed great stock in testing drugs in living systems (in vivo) and was prepared to continue working with a compound even after it failed testing on bacteria cultured in laboratory glassware (in vitro). Among the hundreds of chemical compounds prepared by Mietsch and Klarer for Domagk to test were some related to the azo dyes. They had the characteristic -N=N- coupling, but one of the hydrogens attached to nitrogen had been replaced by a sulfonamide group. In 1931 the two chemists presented a compound (KL 695) that, although it proved inactive in vitro, was weakly active in laboratory mice infected with streptococcus. The chemists made

■ Gerhard Domagk. Courtesy Bayer Corporation.

65

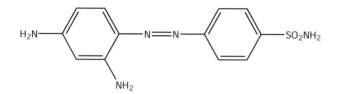

**Figure.** Prontosil.

substitutions in the structure of this molecule and, several months and thirty-five compounds later, produced KL 730 (2´-4´-diamino-azobenzene-4-sulfonamide), which showed incredible antibacterial effects on diseased laboratory mice. It was named prontosil rubrum and patented as Prontosil (Figure).

Domagk spent the next three years investigating the antibacterial properties of Prontosil. He finally published the first report of his findings, "Ein Beitrag zur Chemotherapie der bakteriellen Infektionen," in 1935. In the intervening three years Prontosil had been successfully used to treat several diseases in humans, of both streptococcal and staphylococcal origins, among them Domagk's own six-year-old daughter Hildegard. She had contracted a severe streptococcal infection from an unsterilized needle. She recovered but suffered a permanent reddish discoloration of her skin.

Domagk's discovery of the antibacterial properties of Prontosil won him the 1939 Nobel Prize in physiology or medicine. However, the Nobel committee had angered the German political authorities by awarding the 1935 Nobel Peace Prize to Carl von Ossietzky. Under the grip of Hitler and the Nazi Party, German citizens were forbidden to accept the Nobel Prize. After Domagk accepted the prize, he was arrested by the Gestapo and forced to send a letter rejecting it. Although Domagk was able to receive his prize medal in 1947, the prize money had long since been redistributed.

In their day sulfa drugs developed in Germany and elsewhere were used to good effect to treat, among other conditions, meningitis, childbed fever, pneumonia, blood poisoning, gonorrhea, burns from gas warfare, and other serious burns. But the advent of penicillin during World War II, followed by a host of other antibiotics that were more effective against bacteria, shifted the focus away from Prontosil and the sulfa drugs. Still, bacteria could become resistant to antibiotics, and in the latter part of his career Domagk turned his attention to the search for an antitubercular drug to replace the already increasingly ineffective streptomycin. Although Domagk and his research team were unsuccessful in finding that replacement, their work contributed to the later discovery of isoniazid—one of the strongest and most reliable antitubercular drugs.

■ Gerhard Domagk with microscope. Courtesy Bayer Corporation.

# DANIEL BOVET (1907–1992)

Gerhard Domagk's discovery of Prontosil and its action on streptococcal diseases, published in early 1935, caused other research groups looking for new chemotherapies to redouble their efforts. This was especially true of the group with which Daniel Bovet was working at the Pasteur Institute in Paris. In short order Bovet and his colleagues determined which part of the molecule that the German group had synthesized was responsible for its antibacterial activity. They thus helped open the door for themselves and others to synthesize a panoply of sulfa drugs to treat a variety of illnesses. Bovet was awarded the Nobel Prize in physiology or medicine in 1957 for work in the seemingly quite different field of substances acting on the vascular system and on the muscles. (This field was opened up in the first decades of the twentieth century by Sir Henry Hallett Dale and Otto Loewi; see p. 59.)

Bovet was born in 1907 in Neuchâtel, in the French-speaking part of Switzerland. His father was a psychologist and a cofounder of the child study institute at the University of Geneva, later directed by Jean Piaget. He was educated in Neuchâtel and at the University of Geneva, where he majored in the natural sciences. He then pursued graduate work in physiology and zoology at the university, earning his doctorate in 1929.

In the same year he joined the Pasteur Institute's Laboratory of Therapeutic Chemistry at the request of its director, Ernest Fourneau, with the charge of setting up a pharmacological unit there. Among Bovet's colleagues was the bacteriologist Federico Nitti, son of Francesco Saverio Nitti, a former prime minister of Italy (1919–1920), who was living in exile in France during the Fascist era. In 1938 Bovet married Federico's sister Filomena, who was also a scientist and who collaborated closely and continuously with her husband throughout his career.

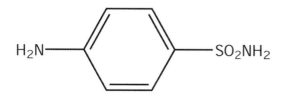

**Figure.** Sulfanilamide.

In reacting to the announcement of Prontosil, Bovet and his colleagues Federico Nitti and Jacques and Térèse Tréfouel prepared and experimented with various products that were related to Prontosil (see Figure, p. 66) but had quite different substituents attached to a nitrogen of the azo bond (-N=N-), the one not attached to the sulfonamide. They observed that a number of compounds whose physical and chemical characteristics differed greatly from Prontosil had the same effect on streptococcal infections in mice and rabbits. Bovet's group then deduced that a number of changes to the molecules must have occurred within the living organism, probably beginning with the breaking of the azo bond, leaving the common constituent of sulfanilamide (Figure). So they started experimenting with it and discovered that this relatively simple colorless substance was the source of Prontosil's therapeutic powers. Domagk and his colleagues, working more directly in Paul Ehrlich's tradition, may have been predisposed to believe that the active part of the molecule was the staining agent. As Bovet noted afterward, the "red car" had a "white engine." Since sulfanilamide had already been synthesized, it was not patentable, so it immediately went into cheap mass-production. Furthermore, Bovet's preliminary experiments with the Prontosil-like molecules gave a better understanding of the relation between structure and function in this class of drugs and supported the development of other sulfa drugs.

By the time sulfanilamide was first discovered, Bovet had already made his mark at the Pasteur Institute as a highly innovative researcher. When he arrived, a search, like the one led at IG Farben by Fritz Mietzsch (see Gerhard Domagk, p. 65), was being conducted for synthetic antimalarials to replace quinine. Bovet's work resulted in the production of Rhodoquine. A by-product of this search was his synthesis of the first synthetic substance to inhibit the activity of adrenaline in the body, Prosympal (1933), which found only temporary use in contemporary medicine. More important, in testing substances in other chemical series, he gained a working understanding of how a molecule must be structured to act like adrenaline in the body or to oppose its activity.

Knowing that there were natural substances that antagonize, or counter the effects of, acetylcholine and adrenaline,

■ Daniel Bovet. Courtesy Istituto Superiore di Sanita.

and now that he had a synthetic antagonist for adrenaline, Bovet set out with his colleague Anne Marie Staub to find a synthetic antihistamine. Their search was rewarded in 1937 by thymoxyethyldiethylamine, which was too toxic to be used as a drug. It did, however, irrefutably show that histamines are most responsible for allergic reactions, including the serious medical condition of anaphylactic shock—a connection that was uncertain up to that time. The antihistamines used today to combat allergic symptoms are derived from this research.

Bovet also followed up Otto Loewi's (p. 59) work on antagonists for acetylcholine, a transmitter substance at many nerve endings, including the junctions between nerves and muscles. Bovet and his associates developed a synthetic curare, the poison used on darts by some South American Indians. The scientists introduced it and other similar agents into surgical practice to relax the body's muscles.

In the era before nerve receptors were even discovered, let alone their molecular structures determined, Bovet and Filomena Bovet Nitti systematized the knowledge gained from all the experimentation sketched above into a bible of molecules discovered to have effects on the sympathetic nervous system: *Structure et activité pharmacodynamique des médicaments du système nerveux végétative* (1948). This book was commonly consulted by a generation of researchers to find the biological activities of various chemical structures on the autonomic nervous system.

In 1947 Bovet accepted an invitation to take charge of a new pharmacological laboratory at the Istituto Superiore di Sanita in Rome. There his new research team continued the work on curare and on a vast range of substances that affect the central nervous system, which he noticed was affected by many of the substances that worked on the autonomic nervous system. In Paul Charpentier's (p. 118) hands this research led a few years later to chlorpromazine, one of the first of the modern drugs used to relieve serious psychological disorders.

In the 1960s, during a period of troubles at the Istituto Superiore di Sanita, Bovet left to take positions at the University of Sassari in Sardinia and at the University of California in Los Angeles. He was particularly happy to come to the United States because of the availability there of inbred strains of mice he could use to pursue his new interest in the genetic basis of learning and memory and the drugs that interfere with memory consolidation.

In 1969 Bovet returned to Rome to become director of a new laboratory of psychobiology and psychopharmacology at the National Research Council, while simultaneously holding a professorship at the University of Rome.

# PART II.

# WORLD WAR II
## TO
## 1960

**Clockwise from top left:** Gladys Hobby and Alexander Fleming (courtesy Pfizer, Inc.);
Percy Julian (courtesy DePauw University Archives and Special Collections); Frances
Kelsey and President John F. Kennedy (courtesy FDA History Office, Rockville, Mary-
land); Albert Sabin (courtesy Hauck Center for the Sabin Archives, Cincinnati Medical
Heritage Center); Dorothy Hodgkin (courtesy Hodgkin family and Judith A. K. Howard);
Selman Waksman (left) and colleagues (courtesy Merck Archives, Merck & Co., Inc.).

# 6. Antibiotics

Antibiotics, introduced in the 1940s, produced miraculous cures. In many well-publicized cases, patients at death's door were brought back to life. The effects were all the more dramatic because in the early phases of development only those patients who did not respond to other therapies—often the sulfonamides—and whose situations were desperate were given the experimental drugs. Some of the most serious afflictions that had previously resisted medical intervention or had been only partially controlled by medication (such as syphilis and tuberculosis) became curable diseases. Because of the introduction of antibiotics during World War II, for the first time in the history of warfare, enemy fire took more lives than disease did.

In the antibiotic era, more than ever before, pharmaceutical research and development drew on knowledge and expertise from a wide range of life sciences, which required a team approach—often teams that connected university or government research laboratories with the pharmaceutical industry. In the penicillin story in particular, the special conditions and demands of wartime vastly expanded the number and variety of professionals involved. Perhaps because of the very success of antibiotics, the teams were often fractured by disputes about who should be honored for various contributions and who could legitimately obtain patents and benefit from royalties.

Despite the miracle-like success of the early antibiotics described in the following pages, finding drugs to fight bacterial infections remained an important research area. Semisynthetic and synthetic antibiotics grew out of the research on penicillin, and this new generation of antibiotics had fewer side effects and was easier to administer.

Today, because most common antibiotics are overused, growing numbers of bacteria have developed resistance to the drugs, in a process similar to natural selection. When bacteria are exposed to antibiotics, most are killed, but the least vulnerable survive. Those bacteria reproduce, passing their survival characteristics on to their offspring. Some bacteria are resistant to multiple drugs, making infections that were once simple to treat life threatening, particularly for children. This plight has increased the interest in antibiotics by pharmaceutical companies. New antibiotics are being developed that can target the resistant characteristics of bacteria, making it possible to treat antibiotic-resistant bacterial infections successfully.

# ALEXANDER FLEMING (1881–1955)

As far back as the nineteenth century, antagonism between certain bacteria and molds, including *Penicillium*, had been observed, and a name was given to this phenomenon—*antibiosis*—but little was made of these observations. A folk tradition using molds in medicine was similarly neglected. In 1928 Alexander Fleming discovered penicillin, but he did not receive the Nobel Prize in physiology or medicine for his discovery until 1945. Fleming himself did not realize how important his discovery was; for a decade after, he focused instead on penicillin's potential use as a topical antiseptic for wounds and surface infections and as a means of isolating certain bacteria in laboratory cultures. It was left to his fellow Nobelists, Howard Florey and Ernst Chain (p. 76), to demonstrate in 1940 that penicillin could be used as a therapeutic agent to fight a large number of bacterial diseases wherever they occurred in the human body.

Born in Lochfield, Ayrshire, Scotland, Fleming was the seventh of eight surviving children in a farm family. His father died when Alexander was seven years old, leaving Fleming's mother to manage the farm with her eldest stepson. Fleming, having acquired a good basic education in local schools, followed a stepbrother, already a practicing physician, to London when he was thirteen. He spent his teenaged years attending classes at Regent Street Polytechnic, working as a shipping clerk, and serving briefly in the army during the Boer War (1899–1902), although he did not see combat. Then in 1901 he won a scholarship to St. Mary's Hospital Medical School, Paddington, which remained his professional home for the rest of his life.

Fleming accepted a post as a medical bacteriologist at St. Mary's after completing his studies, and in 1906 he joined the staff of the Inoculation Department under the direction of Sir Almroth Wright. Wright strongly believed in strengthening the body's own immune system through vaccine therapy, not by chemotherapy. Nonetheless he turned over to Fleming samples of salvarsan that Paul Ehrlich (p. 17) had shipped to the department for testing in 1909. Fleming's experience administering the new drug to patients was positive, and thereafter he maintained a small but lucrative practice administering salvarsan to wealthy patients suffering from syphilis. During World War I, Fleming worked at a special wound-research laboratory in Boulogne, France,

headed by Wright. In France, Fleming began research that produced results more in keeping with Wright's thinking. He was able to demonstrate that then commonly used chemical antiseptics like carbolic acid do not sterilize jagged wounds; rather, pus has its own antibacterial powers. Using cells on a slide, he was able to show that chemical antiseptics in dilutions harmless to bacteria actually damage white blood corpuscles (leukocytes)—the body's first line of defense.

After World War I, Fleming continued to work on leukocytes and antisepsis. In 1921 Fleming discovered a substance in nasal mucus that causes bacteria to disintegrate. He named this substance lysozyme (from the Greek *lysis*, to loosen, and *zymos*, leaven or ferment, because it seemed to act like other known enzymes). Fleming, in collaboration with V. D. Allison, subsequently detected lysozyme in human blood serum, tears, saliva, milk, and a wide variety of other fluids. Lysozyme, in its natural state, seemed to be more effective against harmless airborne bacteria than against disease-causing bacteria. And attempts to concentrate it, thereby strengthening its antiseptic properties, proved unsuccessful.

Fleming's legendary discovery of penicillin occurred in 1928, while he was investigating staphylococcus, a common bacteria that causes boils but can also cause disastrous infections in patients with weakened immune systems. Before Fleming left for a two-week vacation, a petri dish containing a staphylococcus culture was left on a lab bench and never placed in the incubator as intended. Somehow, in preparing the culture, a *Penicillium* mold spore had been accidentally introduced into the medium—perhaps coming in through an open window, as Fleming later supposed, or more likely floating up a stairwell from the laboratory below in which various molds were being cultured. The temperature conditions that prevailed during Fleming's fourteen-day absence permitted both the bacteria and the mold spores to grow; had the incubator been used, only the bacteria could have grown.

Fleming's laboratory notebooks are sketchy, and his subsequent accounts of the discovery are contradictory. The evidence of the first culture, which he photographed (and which still exists, though desiccated) indicated that Fleming

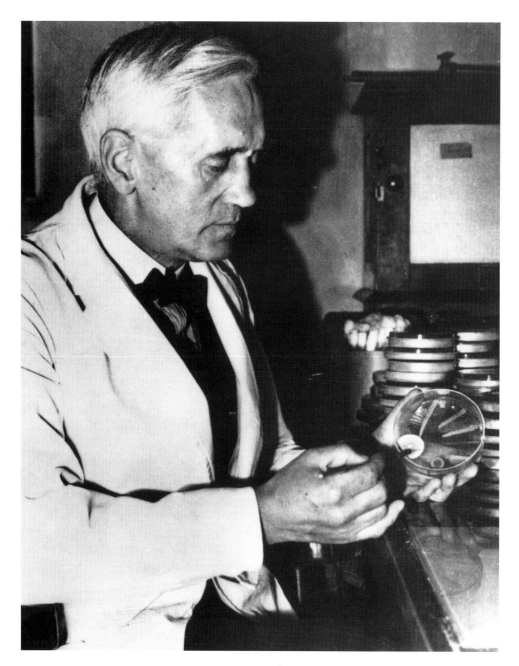

■ Alexander Fleming, holding a petri dish.
Courtesy Bristol-Myers Squibb Corporation.

observed *lysis*, the weakening and destruction of bacteria—as in his lysozyme studies. But sometimes Fleming described the key observation as an instance of inhibition, or prevention of bacterial growth in areas affected by the mold "juice" evidenced by a clear zone surrounding the mold. Although these two effects occur under quite different conditions, Fleming probably forgot which observation came first, for in the months subsequent to the original observation he conducted many experiments while varying conditions systematically.

He discovered that the antibacterial substance was not produced by all molds, only by certain strains of *Penicillium* (originally misidentified by a colleague as *Penicillium rubrum* and later correctly identified as *Penicillium notatum*). Although he could not isolate it, he named the active substance "penicillin." He studied methods of producing the impure product and determined its stability at different temperatures and over various lengths of time. He investigated its effect on many microbes, curiously omitting the familiar spirochete that causes syphilis (which salvarsan controlled but did not eliminate). He tested its toxicity on a laboratory mouse and a rabbit. Forever after, it has been a puzzle why he did not inject these or other laboratory animals with staphylococcus or other disease-causing bacteria before injecting them with the fluid containing penicillin. Perhaps the explanation for his failure to proceed in this direction was his belief that cures come from within the body itself rather than from an external chemical agent. In later years he claimed that the difficulties he had experienced in isolating and stabilizing penicillin, let alone the problems of producing sufficient quantities for clinical trials, had prevented him from realizing the full fruits of his research.

In fact, in the 1930s, little notice was taken by the scientific community of his paper published in the *British Journal of Experimental Pathology* (June 1929). Those few scientists who sent for samples and tried to gain more understanding of the properties of penicillin did not or could not capitalize on Fleming's discovery.

# HOWARD WALTER FLOREY (1898–1968) AND ERNST BORIS CHAIN (1906–1979)

Howard Walter Florey and Ernst Boris Chain, the scientists who followed up most successfully on Fleming's discovery of penicillin, each brought scientific knowledge and talent to the effort that filled out the other's contribution, but the two were mismatched in terms of their personalities.

Florey was born in 1898 in Adelaide, Australia—the youngest of five children and the only son of an English shoemaker who had immigrated to Australia hoping to save his first wife and two eldest daughters, who were suffering from tuberculosis. He established a boot and shoe factory, which prospered during much of Florey's youth.

From an early age Florey knew that he wanted to study medicine, perhaps because an older sister was already a medical student, at a time when few women became physicians. She teased him about wanting to become "another Pasteur," thus forecasting his interest in medical research. In 1922 Florey graduated with degrees in science and medicine from the University of Adelaide, which at that time specialized in preparing general practitioners, not researchers. On the basis of his fine academic record and his prowess as a tennis player, he was awarded a prestigious Rhodes Scholarship to study at Oxford University in England.

At Oxford his hunger for research was fed at the Honours Physiology School, which was headed by the great neurophysiologist Sir Charles Sherrington. After completing his Rhodes Scholarship, Florey spent a summer as the physician on an Arctic expedition and the following year as a research student at Cambridge. He then accepted a fellowship from the Rockefeller Foundation in New York that enabled him to work in Alfred Newton Richards's laboratory at the University of Pennsylvania. Soon after he returned to England, he began working toward a Ph.D. in pathology at Cambridge University. There he was inspired by the biochemist Sir Frederic Gowland Hopkins, head of the Sir William Dunn Institute of Biochemistry and famous for his work on vitamins. Florey also worked with Albert Szent-Györgyi (p. 30), who was well along in his work isolating vitamin C.

After holding the first few positions in his career as an academic, he returned to Oxford in 1936 as the director of the Sir William Dunn School of Pathology. There he re-

cruited an interdisciplinary group of scientists, a choice that reflected the various approaches to disease to which his education had exposed him—studies not just of the pathological evidence of disease but also of the physiological processes by which those symptoms arose, traced to the chemical and even the molecular level. By 1940 Florey had assembled a professional staff of about a dozen scientists plus technicians to work on the penicillin project alone. Among his first hires was the biochemist Ernst Boris Chain, who came highly recommended by Frederick Hopkins at Cambridge and who was charged with developing a biochemical unit within the Dunn School at Oxford.

Chain, a recent immigrant to England, was of Russian-German-Jewish descent. He was quite confident in his own ability and had a volatile temperament that was bound to clash with Florey's similarly quick temper. Chain had left his native Berlin in 1933, as soon as the Nazis came to power, leaving behind his mother and sister (who would both perish in the Holocaust). When Chain was only thirteen, his father, an industrial chemist, died. The family's economic status declined precipitously, causing his mother to convert their home into a guest house. The father's influence lived on in the son, however, and Chain chose a scientific course of studies, graduating in chemistry and physiology from Friedrich Wilhelm University in Berlin in 1930. After completing his degree, he worked at premier German research institutes—briefly at the Kaiser Wilhelm Institute for Physical Chemistry and Electrochemistry (at that time still under the direction of Fritz Haber) and then in the Chemical Department of the Pathology Institute at Charité Hospital in Berlin, where he obtained his doctorate. From childhood Chain's musical gifts were comparable to his scientific genius, and at virtually every career decision point through his twenties—even after he came to England—he entertained the possibility of becoming a professional pianist instead of a scientist.

The first job offer in response to his many letters of application in his new country came from Charles Harington of the chemical pathology department at University College Hospital Medical School, London, who would later synthesize the hormone thyroxin. Chain took the position but quickly alienated himself from the rest of the depart-

■ Howard Florey in 1945. Courtesy Mary Evans Picture Library.

ment, partly through his endless complaints about the low quality of laboratory equipment in comparison with the situations he had known in Germany. Chain worked at Cambridge under Florey's mentor Hopkins from 1933 to 1935, then moved to the William Dunn School at Oxford.

With Chain aboard, one of the projects pursued at the school was the crystallization of lysozyme and the characterization of its substrate—the location on bacteria to which it usually attaches. In 1938, while the lysozyme research was concluding and during a rare period of great camaraderie, Florey and Chain decided to study the biochemical and biological properties of selected antibacterial substances produced by certain microorganisms. They mistakenly thought these substances were all enzymes like lysozyme. This process was greatly facilitated by Chain's near-photographic recall of the many scientific papers he had read, including Fleming's 1929 paper on penicillin. They originally chose substances from three organisms: pyocyanase, a topical antibacterial, from *Bacillus pyocyaneus*; extracts from certain organisms in the soil called actinomycetes; and penicillin.

While Florey and Chain were assembling grants from Britain's Medical Research Council and the Rockefeller Foundation to support their research program, work was begun on penicillin. Fortuitously, there was already a penicillin culture at the William Dunn School. This culture had been kept growing since the days of Florey's predecessor, who had obtained it from Fleming with the thought that it might have some connection with his research on bacteriophages, viruses that have a special affinity for bacteria (which it did not). The research program rapidly narrowed its focus to penicillin alone, although one worker in the laboratory was deployed to work on pyocyanase. The immediate future of actinomycetes research lay in the United States with René Dubos, at the Rockefeller Institute, and Selman Waksman (p. 90), at Rutgers University.

Chain, along with another chemist, Edward Penley Abraham, worked out a successful technique for purifying and concentrating penicillin. The keys seemed to lie in controlling the pH of the "juice," reducing the sample's temperature, and evaporating the product over and over (essentially freeze-drying it). In this early process gallons and gallons of mold broth were used to produce an amount just large

enough to cover a fingernail—an excruciatingly inefficient process that was later improved on by Norman Heatley and a succession of other scientists.

In March 1940 Chain ran down to a laboratory that maintained test animals and requested that two mice be injected with a sample of the penicillin that he and Abraham had extracted. Though the injection represented a far higher dosage than that administered in Fleming's similar experiment, the mice survived apparently unharmed; the more-concentrated penicillin had passed its first toxicity test. Florey then directed that the antibacterial properties of penicillin in mice be tested—the step that Fleming had not taken. On 25 May eight mice were injected with hemolytic streptococci (which among other diseases causes puerperal fever in new mothers), and four of these were subsequently injected with measured and timed doses of penicillin. Sixteen-and-one-half hours later the four mice that had received penicillin were alive (with only one seeming to drag a bit), but their untreated fellows were dead—a finding that caused great excitement in the group. Further testing involving hundreds of mice was carried out through the summer. On 24 August 1940 Florey and Chain reported their findings in the *Lancet*; the article electrified research groups around the world that were seeking cures for bacterial disease. By then World War II had already engulfed Europe, and the military importance of a more successful means of combating the diseases and infections that had decimated armies of the past was immediately recognized.

In early January 1941 Florey was ready to test penicillin on humans. The first English patient to whom the drug was administered was a young woman whose cancer was beyond treatment and who had agreed to test penicillin's toxicity. She showed an alarming reaction—trembling and sharply rising fever. However, using the relatively new technique of paper chromatography, Abraham was able to separate out the impurities from the penicillin. He was then able to show that it was these impurities, not the drug itself, that had caused the adverse reaction. On 12 February 1941 a policeman suffering from an invasive infection that had begun with a simple thorn scratch on his cheek became the first patient with an infection to be treated with penicillin in the hope of achieving a cure. No one knew the dosages and the

■ Ernst Chain in 1945. Courtesy Mary Evans Picture Library.

length of treatment required to eliminate various bacterial infections; these parameters were being worked out by just such trials—primitive by today's standards. The policeman's condition at first improved with the penicillin therapy and then relapsed. The penicillin supply had almost run out, and even retrieving penicillin from the man's own urine (a commonly used procedure in the early clinical trials) failed to save him. Florey vowed that from then on he would always have enough penicillin to complete a treatment.

Increasing production and yields—problems that had been addressed to some degree when testing began on mice—now became of overriding importance. Because *Penicillium* mold requires air to grow, it was first surface-cultured in regular laboratory flasks. Soon all manner of vessels were being used, including hospital bedpans and hundreds of made-to-order ceramic pots. The operation quickly outgrew the space assigned to the William Dunn labs, and neighboring facilities at Oxford were borrowed for the duration. More personnel had to be hired, including six "penicillin girls" who handled the culture pots in the cold room of the extraction plant. Florey had constructed a veritable penicillin factory within the precincts of the ancient university, an institution that had stood proudly aloof from industry for centuries. On the other hand, when Chain urged that a patent be sought on penicillin, as was usual in German research institutes, Florey refused to enter into such a commercial agreement on a discovery he presumed would benefit all mankind—a decision that long rankled Chain.

To increase penicillin supplies, Florey approached various British pharmaceutical firms, but only ICI considered itself in a position to accept the challenge (though many later joined the effort). British pharmaceutical firms were already committed to manufacturing other drugs needed for military and civilian populations, or, worse, their facilities had been devastated by enemy bombardment. To obtain the assistance of the United States, then still a noncombatant, in increasing production and furthering research, Florey and Heatley flew across the Atlantic in the beginning of July 1941.

Florey's American connections served him well. The two English emissaries spent the Fourth of July weekend with his friend from his Rhodes year, John Fulton, and his wife, who had been caring for the two Florey children for some time to remove them from the terrible German bombardment of England. Fulton made the phone call that eventually put Florey and Heatley in contact with the U.S. agricultural research establishment, where large-scale fermentation processes were being actively studied. A. N. Richards, Florey's old laboratory director, had become chair of the Committee on Medical Research in the Office of Scientific Research and Development, organized to marshal the strength of the Allies. Because Richards knew Florey's character, he decided to expedite unified action on penicillin on the basis of just one presentation. At the height of the program the British-American penicillin effort involved thousands of people and some thirty-five institutions: university chemistry and physics departments, government agencies, research foundations, and pharmaceutical companies.

In mid-September, Florey flew back to England to lead the clinical trials that eventually took him as far afield as North Africa and the Soviet Union (then one of the Allies). Heatley stayed at the Northern Regional Research Laboratories of the Department of Agriculture in Peoria, Illinois, and later moved to Merck and Company in Rahway, New Jersey, to help iron out problems in scaling up production.

Some chemists were confident that they would soon be able to synthesize penicillin from a few organic chemicals—as they had salvarsan and the sulfonamides. This attitude resulted in a major effort conducted on both sides of the Atlantic to understand the structure of the penicillin molecule as the prerequisite for its eventual synthesis. In those days such work usually meant reacting the substance under study with various chemical reagents, which resulted in products of known structure. From these bits of structural information scientists could deduce how the original molecule was organized. But often more than one plausible interpretation could be made. At Oxford the problem of determining penicillin's structure was given to Chain, Abraham, and Robert Robinson, a senior organic chemist already famous for solving the structures and the synthesis of many complex molecules derived from natural substances.

By fall 1943, groups working at Oxford and at Merck had proposed two different structures for the atomic groups central to the penicillin molecule. One proposal—advanced by Chain and Abraham as well as by Robert Burns Wood-ward (p. 114) at Harvard—held that a four-membered beta-lactam ring lay at the heart of the penicillin molecule (Figure 1). Robinson instead proposed a structure based on oxazalone (Figure 2). Meanwhile new instrumental techniques for analyzing the structure of organic molecules had become available, including X-ray crystallography, which was practiced by Dorothy Hodgkin, a near neighbor of the Oxford chemists (p. 82). In 1945 she was able to provide physical evidence confirming Chain and Abraham's deduction. This evidence ran counter to Robinson's proposed structure for penicillin. Unfortunately, Hodgkin's work on the molecule culminated too late in the war to be used in devising a synthesis. Even after 1957, when John Sheehan created such a synthesis, fermentation continued to underlie the commercial production of penicillin and related antibiotics. But the structural knowledge gained in the war years proved invaluable in developing penicillin-like antibiotics after the war that could be administered more conveniently, were more effective, and had fewer side effects.

With World War II ended and the Nobel Prizes in physiology or medicine distributed to Fleming, Florey, and Chain for their work on penicillin, Florey continued to lead the William Dunn laboratory along the promising path of research into antibiotics. One of his group's most famous accomplishments was the development of cephalosporin C in 1954. In this case Florey immediately urged its patenting.

In 1962 Florey became provost of Queen's College, Oxford, and from 1960 to 1965 he held one of the world's

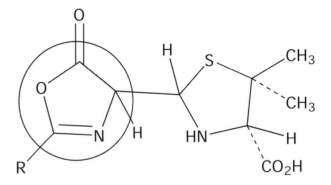

**Figure 2.** Proposed oxazolone structure for penicillin.

most prestigious scientific positions, the presidency of the Royal Society. To remedy the deficiencies of Australian higher education as he had experienced it, he helped organize the Australian National University in Canberra as a postgraduate-level university, and in 1965 he became its chancellor.

After World War II, Chain was eager to leave Oxford. Of the several career moves he considered, he ultimately chose the Istituto Superiore di Sanita in Rome, the same institute to which the Frenchman Daniel Bovet (p. 68) came after the war. In Rome, Chain productively combined a biochemical research department and a fermentation pilot plant. In 1957 a consulting relationship with a group of scientists from the Beecham Group, who came to Rome especially to benefit from Chain's biochemical insights and the facilities there, resulted in the isolation of the atomic groupings central to the penicillin molecule. Knowing how to produce this "nucleus" by natural processes, researchers could move on to the commercial production of all kinds of semisynthetic penicillins, such as penicillin V and ampicillin.

In 1964 Chain returned to England to head a new biochemistry department at Imperial College, London, which he specified would include a fermentation pilot plant as one of its facilities. Always a person of many interests and projects, in this later period of his life, Chain made considerable efforts to promote respect for Jewish traditions.

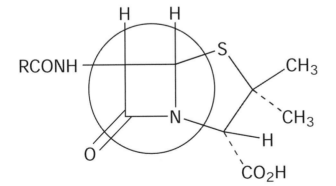

**Figure 1.** Beta-lactam ring structure for penicillin.

# DOROTHY CROWFOOT HODGKIN (1910–1994)

On the morning in May 1940 that the results of injecting four of eight infected mice with penicillin became known, an excited Ernst Chain encountered Dorothy Hodgkin on Parks Road in Oxford. He knew the brilliant young X-ray crystallographer from her Cambridge days, and as she later recalled, he had promised: "Some day we will have some crystals for you to work on."

Dorothy Crowfoot was born in Cairo, Egypt, to English parents. Although her formal schooling took place in England, she spent a significant part of her youth in the Middle East and North Africa, where her father was a school inspector. Both her parents were authorities in archaeology, and she almost followed the family vocation; but from childhood she was fascinated by minerals and crystals. She enjoyed using a portable mineral analysis kit given to her when she became interested in analyzing pebbles she and her sister found in the stream that ran through the Crowfoots' garden in Khartoum, Sudan.

When she was fifteen, her mother gave her *Concerning the Nature of Things* (1925), a popular book by one of the founders of X-ray crystallography, Sir William Henry Bragg. It contained intriguing discussions of how scientists could use X-rays to "see" atoms and molecules. In this method structural information is obtained from the substance under observation by mathematical analysis of the intensity of X-rays scattered (or diffracted) from parallel planes in a crystal, as recorded photographically or by an electronic detector.

At Somerville College, Oxford, Crowfoot studied physics and chemistry and chose to do her fourth-year research project on X-ray crystallography, a field for which the equipment at Oxford was primitive. After graduation she seized the opportunity of studying at Cambridge with John Desmond Bernal, who had worked for five years with Bragg at the Royal Institution. Crowfoot and Bernal collaborated successfully, using X-ray crystallography to elucidate the three-dimensional structure of complex and biologically important molecules, including pepsin—the digestive enzyme found in the stomach that was the first protein to be

so analyzed. In 1937 Crowfoot received her Ph.D. from Cambridge—the same year she married Thomas L. Hodgkin, a historian who became an authority on Africa. Both Hodgkins held academic appointments at Oxford, and they raised their three children there with the help of the Hodgkin grandparents.

The structural determination of penicillin, one of Hodgkin's greatest chemical achievements, depended on the first step of getting the substance to crystallize (as do all such determinations). The strain of penicillin with which the Americans were working proved easier to crystallize, and a three-milligram sample was flown across the Atlantic for Hodgkin to crystallize personally. To perform the thousands of calculations necessary to transform the two-dimensional X-ray data into three-dimensional atomic positions, the group under Hodgkin's leadership made use of a cast-off punched-card computer, and in 1945 they were able to confirm the beta-lactam structure central to the penicillin molecule (see Howard Florey and Ernst Chain, Figure 1, p. 81).

For the determination achieved in 1957 of the structure of vitamin $B_{12}$, the vitamin essential to preventing pernicious anemia, Hodgkin's group started with the old punched-card machine, but in time they used electronic computers in England and in the United States. In 1964 she was awarded the Nobel Prize in chemistry for her work on vitamin $B_{12}$, the most complex organic molecular structure determined in detail up to that time. In 1969 she completed a three-dimensional structure of insulin—the hormone essential for the control of blood sugar levels—following Frederick Sanger's ordering of insulin's fifty-one amino acids, which brought him the 1958 Nobel Prize in chemistry (see Frederick Grant Banting et al., p. 52).

Hodgkin is fondly remembered by her group of research students, which included many women. She was also involved in a wide range of peace and humanitarian causes. From 1976 to 1988 she was chair of the Pugwash movement, which elicits the insights of and input from the world's scientists on potential dangers raised by scientific research.

■ Dorothy Hodgkin in her early days at Somerville College, in Oxford, England, 1930s. Courtesy Hodgkin Family and copy from Judith A. K. Howard (Durham, U.K.).

■ Dorothy Mary Crowfoot Hodgkin. Portrait by Maggi Hambling. Oil on canvas, 36.75 inches × 30 inches (932 mm × 760 mm), 1985. This portrait was the first of a woman scientist to be hung in the National Portrait Gallery, London. Courtesy National Portrait Gallery, London.

# Gladys L. Hobby (1911–1993)

Among the researchers who reacted to the August 1940 *Lancet* article written by Howard Florey and Ernst Chain (p. 76) was Gladys Hobby, a microbiologist and later a historian of penicillin. Hobby was part of a group of scientists at Columbia University's College of Physicians and Surgeons that was studying hemolytic streptococci and the diseases they produce, especially subacute bacterial endocarditis, an infection of the heart valves that often followed rheumatic fever and had always proved fatal. The group also included Martin Henry Dawson, clinician and leader, and Karl Meyer, chemist. Recognizing the potential use of penicillin against hemolytic streptococci, the group immediately sent for a subculture from Oxford as well as one from Roger Reid, who had worked with a culture in the 1930s shortly after Fleming made his discovery.

Born in Manhattan, Gladys Hobby graduated from Vassar College in 1931 and earned her master's degree and a Ph.D. in bacteriology from Columbia University. In 1934 she joined a small research team in the Department of Medicine at the College of Physicians and Surgeons, where she remained for a decade.

When the cultures arrived, Hobby quickly set to work growing them. And soon, as at the William Dunn School of Pathology, containers filled with the mold with its tiny golden droplets were proliferating in the medical school—until space under a university athletic stadium was commandeered as an ideal place to grow molds. Starting with primitive methods, Meyer was able to extract small, relatively crude quantities of penicillin from the mold broth. On 15 October 1940, less than a month after receiving the cultures and months in advance of the projected schedule (and three months in advance of Florey's group), Dawson administered penicillin to a few patients at Columbia Presbyterian Hospital—at first to test its toxicity. It was not until March 1941 that it was administered in sufficient doses to show therapeutic action. Soon, miraculous cures like the ones already experienced in England were being witnessed.

The first scientific paper from Dawson's group was presented in May 1941 at the annual meeting of the American Society for Clinical Investigation held in Atlantic City. The paper caused quite a stir in the popular press as well as in the scientific world. Among the attendees were Pfizer scientists, and soon the company initiated its own studies of penicillin. By fall 1941 Pfizer was producing enough penicillin to supplement and by 1944 to replace the supply manufactured by Hobby and Meyer at the College of Physicians and Surgeons.

In 1944 Hobby left Columbia to join Pfizer, where her work on antibiotics continued. There her research included the development of penicillin, streptomycin, Terramycin, and other antibiotics. In 1959 she became chief of research at the Veterans Administration Hospital in East Orange, New Jersey, where she studied chronic infectious diseases and remained until her retirement in 1977.

■ Gladys Hobby with Alexander Fleming on one of his visits to the United States. Courtesy Pfizer, Inc.

# ANDREW J. MOYER (1899–1959)

Before the problem of penicillin culture was turned over to the Northern Regional Research Laboratory (NRRL) of the U.S. Department of Agriculture (USDA), providing the mold with the necessary oxygen for its survival and growth always entailed large—and ultimately inefficient—surface areas. Howard Florey (p. 76) and Norman Heatley, at the suggestion of USDA officials with whom they met in Washington, D.C., were invited to the NRRL in Peoria, Illinois, arriving there on 14 July 1941. They shared their knowledge and a sample of penicillin with the staff of the fermentation department in this year-old government institute set up to explore new uses for American agricultural products. While Florey headed off to Canada to interest Canadian firms in the commercial production of penicillin, Heatley remained to work with Andrew J. Moyer. This collaboration turned out to be another productive relationship in the penicillin saga.

Moyer was a graduate of Wabash College in Indiana, his home state, with a master's degree from North Dakota Agricultural College and a Ph.D. in plant pathology from the University of Maryland (1929). He had been working with fermentation processes at least since his doctoral thesis on the growth responses of fungi to boron, manganese, and zinc. At first Heatley found it difficult to get the English mold to start growing again after its long journey, but grow it did. It was Moyer who suggested that the growth medium be enriched with corn steep—the otherwise useless by-product of the wet milling of corn to make cornmeal, which the NRRL was trying out in virtually all its fermentation experiments. The effect was magical; the mold grew as never before. Later, other adjustments were made to the medium—substituting a sugar present in milk for a sugar extracted from corn, adding trace metals and other ingredients. Heatley asked to use the rotary drums and vats in the laboratory, which were the laboratory-scale models of "deep-tank," or submerged, fermentation, whereby oxygen was supplied to aerobic microbes by shaking or stirring the contents.

On 17 December 1941 Moyer's boss, Robert D. Coghill, reported Moyer and Heatley's success in culturing the mold, which had resulted in a twelvefold increase in the yield of penicillin, to a meeting of the Committee on Medical Research held at the University Club in New York. Representatives were there from Pfizer and Company, Merck and Company, E. R. Squibb and Sons, and Lederle Laboratories. It has even been said that the antibiotics industry in the United States was born at that meeting. These companies, which were already producing penicillin by surface fermentation, immediately set to work experimenting with deep-tank methods and the new medium.

By the time of the New York meeting Heatley had already moved to Merck to help them produce penicillin. Meanwhile the NRRL continued its research on penicillin, which spanned several departments at the laboratory. Again, who got credit and who succeeded in getting patents resulted in antagonisms all around. Moyer felt slighted by the press from the beginning, and many British scientists were angered to see Americans, including Moyer, gaining patents on processes to which they felt they had contributed significantly. Nearly thirty years after his death Moyer was recognized by the National (U.S.) Inventors Hall of Fame for his U.S. patents relating to the production of penicillin.

■ Andrew Moyer in the Fermentations Division Laboratory, Northern Regional Research Laboratory, USDA, Peoria, Illinois, in 1941. Courtesy Edgar Fahs Smith Collection, University of Pennsylvania Library.

■ John McKeen with Pfizer products, including penicillin and Terramycin. Courtesy Pfizer, Inc.

# John Elmer McKeen (1903–1978)

On 1 March 1944 Pfizer and Company opened the first commercial plant for large-scale production of penicillin by the submerged-fermentation method that Andrew Moyer had pioneered at the Northern Regional Research Laboratory (see p. 86). Chemical engineer John McKeen, superintendent of Pfizer's Brooklyn works, played a critical role in creating a production facility.

A native of Brooklyn, McKeen graduated from Brooklyn Polytechnic Institute in 1926 and immediately went to work for Pfizer. In the 1920s Pfizer was innovating commercial processes by using fermentation to produce chemicals: citric acid by the action on molasses of enzymes from *Aspergillus niger*, common black bread mold, and gluconic acid by the action on glucose of this same mold, cultivated by deep-tank fermentation. Young McKeen, a process development engineer, was very much a part of that exciting period through his close association with the research and development departments. He was appointed head of one of the manufacturing departments in 1935 and was later posted to England to set up a fermentation department there.

Ever since 1941, when Pfizer began supplying the College of Physicians and Surgeons research group with penicillin, it, like other companies producing penicillin for the war effort, used surface-fermentation methods. In industry many recognized that only deep-tank fermentation methods would be economically feasible, and experimentation on these began in several companies. By early 1943 Pfizer's efforts began to yield results. Because of its strong background in fermentation processes, Pfizer forged ahead more quickly than other companies. By the end of the summer of 1943 a two-thousand-gallon pilot fermentor was up and running. The plant was completed the following March; it contained fourteen seven-thousand-gallon tanks plus equipment for recovery and purification of the penicillin. Some 250 employees per shift were engaged in the plant, which ran day and night. Improvements were continuously made in the productivity of the mold and the efficiency of the extraction process. By the end of 1944 the plant produced 723 "Oxford" units (the strength-quantitative units in which penicillin was then measured) for that year alone, more than a hundred times what Pfizer had been able to produce by surface fermentation in 1943 and nearly half the nation's total production for 1944. By the end of the war Pfizer was also manufacturing Selman Waksman's streptomycin (see p. 90).

McKeen, a zealous, innovative, hardworking man who loved people, was destined to go far at Pfizer. In the years before the war he moved through many departments at the company, learning the intimate details of the processes and improving on them, solving problems, saving money, and increasing efficiency. During the war McKeen spent much of his time on penicillin. Spurred by the shortages of materials brought on by the war, he aggressively hunted down the machinery and pipes needed to make the deep-tank process work. His purchase of an old Brooklyn ice-making plant with refrigeration equipment that could be reused in the penicillin plant was the primary reason that the project took less than five months. Once the new plant was up and running, John L. Smith, then a vice president and later president of Pfizer, donated massive quantities of the drug to New York hospitals. McKeen and Smith frequently spent their weekends visiting the patients who were receiving treatment, seeing for themselves the miracle of penicillin.

After the war McKeen became executive vice president and then president of Pfizer (1949–1965), adding in 1950 the responsibility of chairman of the board. He led Pfizer through the "antibiotic era," which included such triumphs as Terramycin and tetracycline. It was also during this period that Pfizer decided to market its prescription drugs under its own label to retail pharmacies and hospitals. (Up until 1950 Pfizer sold its pharmaceuticals in bulk to other companies.)

# SELMAN ABRAHAM WAKSMAN (1888–1973)

Selman Abraham Waksman chose to title his autobiography *My Life with the Microbes* (1954), an apt characterization of the professional life of this biochemist and microbiologist. Waksman, assisted by his students, mined pharmaceuticals from the soil using methods that have become classic. It was a former student, René Dubos, who in a form of role reversal inspired his mentor Waksman to steer his research in this direction. Dubos had isolated the soil bacterium *Bacillus brevis* in the 1930s, and Waksman, after noting the extreme toxicity of the substances that this bacterium produced, began to suspect that soil bacteria were not the most likely allies in the fight against disease. Waksman, already the author of the standard text in the field, *Principles of Soil Microbiology* (1927), thus applied his in-depth knowledge of *all* classes of soil microbes to this problem. And he assembled a troupe of graduate students under his expert tutelage to perform the required laborious experiments.

Born in a small Jewish village in the Ukraine, Waksman was brought up in a family dominated by women—his grandmother, his mother, and seven aunts. His father fled this matriarchy, spending most of his time in the village prayer house, but he gave his son a good religious upbringing. The son progressed from local religious schools to private tutors in secular subjects and ultimately to state secondary schools. Waksman's early life was shaped in part by the death of his little sister from diphtheria—at a time when the antitoxin was well known but not available in their part of the world. Waksman, a socially responsible person, and his friends organized a school for poor boys and later a health service to care for the sick. Life in Jewish villages became precarious as a result of the Russo-Japanese War (1904–1905) and the accompanying Revolution of 1905, during which Jews were blamed for all misfortunes. In response to the pogroms carried out by the infamous Cossacks, Waksman, while still a teenager, helped to organize a youth self-protection league.

Waksman passed his school's final examinations soon after his mother died, and he then left the Ukraine for the United States, arriving in Philadelphia in 1910. He went to stay with a cousin who was farming in Metuchen, New Jersey. From 1911 to 1915 he attended nearby Rutgers College, majoring in agricultural science at the urging of Jacob G. Lipman, a bacteriologist who was dean of the College of Agriculture and himself an immigrant from Russia. Waksman, turned down as a candidate for graduate study at the University of Illinois, stayed at Rutgers for a master's degree. His research, carried out at the New Jersey Agricultural Experiment Station on the Rutgers campus, encompassed the topics that fascinated him for his lifetime: soil bacteria, actinomycetes (filamentous bacteria), and fungi. He next obtained a Ph.D. in biochemistry in two years from the University of California at Berkeley and returned to Rutgers in 1918 as a lecturer in soil microbiology.

To sustain himself financially through his lengthy education and in his early professional life, he held many jobs, some simultaneously. Among these, he was responsible for the irrigation system of a ranch near Sacramento that grew peas and beans. In California he also worked for Cutter Laboratories, a manufacturer of antitoxins, vaccines, and serums. During his first two years as a lecturer at Rutgers he worked at Takamine Laboratories (see Jokichi Takamine, p. 47) in Clifton, New Jersey, preparing fungal enzymes and performing toxicology tests on salvarsan, which had been licensed to this company by Sahashiro Hata, Paul Ehrlich's Japanese collaborator (see Ehrlich, p. 17). Later, as Waksman rose through the academic ranks, he maintained consultative relationships with many industrial concerns that produced enzymes, vitamins, and other products from fungal and bacterial sources. Another of his interests was the study of marine microbes, which he pursued every summer from 1931 onward at the Oceanographic Institute in Woods Hole, Massachusetts.

In 1939 Waksman began the work that would bring him the most fame. With the help of some fifty graduate students and visiting scholars over a number of decades, he screened thousands of soil microbes. This grand scale of research quickly became the world standard, and so Waksman had many assistants to choose from who were interested in learning his techniques. As many soils as possible were sampled, and from these a wide variety of actinomycetes (as well as other microbes) were isolated and tested for their activity against bacteria. The most promising were grown on various liquid media to find those microbes that had the

■ Selman Waksman (left) with colleagues outside 150,000-gallon deep-tank fermenters for making streptomycin at Merck's Stonewall Plant in Elkton, Virginia. Courtesy Merck Archives, Merck & Co., Inc.

capacity to liberate freely the substances possessing such activity. Then attempts were made to isolate the active substances by using, among other means, various solvents, mechanical means, and temperature gradients.

In 1939 representatives from Merck and Company approached Waksman about setting up a cooperative arrangement to undertake the follow-up research and development necessary to convert laboratory discoveries into pharmaceutical products. Specifically, Merck agreed to undertake research to determine the chemical structure of likely substances, to test for their toxicity and pharmaceutical activity in laboratory animals, and to design and carry out the scaling up of their production, so that clinical trials in animal and human subjects could be done. Merck would also seek patents on behalf of Waksman and his named assistants. If the company should decide to manufacture a substance, then it would be the sole producer, and it would pay Rutgers University a percentage of royalties on the sales of bulk product.

Meanwhile Waksman tried to obtain outside financial support for the activities going on in his Rutgers laboratory. First he approached the Committee on Medical Research, but while penicillin appeared to be a good bet to the committee, sifting through soil samples for pharmaceutically active substances seemed too hypothetical. The committee chairman, A. N. Richards, was more sanguine; he steered Waksman to the Commonwealth Fund, a private foundation that did offer support. Later, Mrs. Albert Lasker, who privately supported a great deal of medical research, provided additional funds.

In short order Waksman's group isolated antimicrobial agents from various soil actinomycetes, but only streptomycin, isolated in 1943 by his student Albert Schatz, proved safe for use on human subjects. Much later the laboratory's first discovery, actinomycin, came to be used as a means of killing cancer cells. In the midst of these early discoveries Waksman was asked by the editor of a scientific journal to give a name to these and other antimicrobial substances. He chose the term *antibiotics* based on *antibiosis*, a word created in the nineteenth century to express the injurious effect of one organism on another; Waksman's usage implied that such activity is solely against microbes that cause diseases in humans or other higher forms of life. During his lifetime he and members of his laboratory discovered many more antibiotics, including two more that reached the market.

As an antibiotic, streptomycin succeeded beyond Waksman's wildest dreams. Its most important triumph was over many forms of tuberculosis. Until the mid-twentieth century tuberculosis was a dreaded killer, much more prevalent than AIDS is today. Tuberculosis affected all age groups but, like AIDS, had particularly devastating effects on young adults (usually the healthiest segment of the population). The clinicians H. Corwin Henshaw and William H. Feldman, working at the Mayo Clinic in Rochester, Minnesota, established streptomycin's curative activity in tuberculosis in guinea pigs, and clinical trials on humans began soon thereafter.

When it became obvious that streptomycin would be a lifesaver, Waksman sought release from the exclusivity clause of the original 1939 agreement, to which Merck readily agreed. The patent was transferred to Rutgers University and licensed to seven pharmaceutical companies, including Merck. In the midst of this restructuring a heated dispute broke out between Albert Schatz and his former mentor, resulting in a court judgment that Schatz, who had been named in the patent, was indeed the codiscoverer of streptomycin. The financial settlement, reflecting Waksman's own views of scientific credit, gave 10 percent to all the Rutgers scientists, students, and technicians who played a role in the search for antibiotics (with just 3 percent of that going to Schatz), 10 percent to Waksman, and 80 percent to the Rutgers Research and Endowment Foundation. At Waksman's request the foundation created the Institute of Microbiology in 1951, and he contributed most of his portion to the support of scientific research. Among many other honors recognizing his research on antibiotics, Waksman (alone) was awarded the 1952 Nobel Prize in physiology or medicine.

# Elizabeth Lee Hazen (1885–1975) and Rachel Fuller Brown (1898–1980)

Today a growing number of people—recipients of organ transplants, burn victims, those on chemotherapy, and AIDS patients—are rendered susceptible to fungal infections. We have become increasingly aware of fungi as real threats to life, as well as being among the causes of common, less serious conditions like "athlete's foot."

One of the first effective antifungal medicines, nystatin (or mycostatin), was discovered in 1950 by Elizabeth Lee Hazen, a microbiologist, and Rachel Fuller Brown, a chemist. Their research at the Division of Laboratories and Research of the New York State Department of Health was inspired by the discovery and development of penicillin and by Selman Waksman's success in screening soil samples for antibacterial agents (see p. 90).

Hazen came to work at the New York City office of this division in the 1930s, soon after completing her doctoral degree in microbiology at Columbia University's College of Physicians and Surgeons. Her career path was long and tortuous before her arrival at this model public-health institution famed for fostering a spirit of free inquiry and welcoming women scientists—a rarity in the early twentieth century.

Hazen was born in 1885 in Rich, Mississippi, sixty miles south of Memphis, Tennessee. Orphaned by the age of three, she and her two siblings lived first with their maternal grandmother and then with their father's brother, who had three of his own children. After attending a one-room school, she enrolled in the college known today as Mississippi University for Women. She then taught high school physics and biology for a time in Jackson, Mississippi. During her teaching days she enrolled in summer study at the University of Tennessee and at the University of Virginia. Later she moved to New York City to attend Columbia University, where, despite initial doubts cast by university authorities on her Southern educational background, she completed a master's degree in biology. She embarked on a program in medical bacteriology at Columbia's College of Physicians and Surgeons, which she interrupted to serve in World War I in army diagnostic laboratories in Alabama and New York. After the war she worked as assistant director of the clinical laboratory of a hospital in West Virginia. In 1923 she returned to the College of Physicians and Surgeons and com-

pleted her Ph.D. in microbiology four years later. She then secured clinical and teaching positions at Columbia Presbyterian Hospital and at the College of Physicians and Surgeons.

In 1931 Hazen was lured to the New York office of the Division of Laboratories and Research of the State Department of Public Health and soon was put in charge of its Bacterial Diagnosis Laboratory. In 1944 Augustus Wadsworth, founder and head of the division, chose her to lead an investigation into fungi, which Waksman and others had recognized were in an age-old competition with bacteria and other microbes. Hazen then studied mycology at the College of Physicians and Surgeons and set up a collection of disease-causing fungi housed at the division, which were used to identify microorganisms in specimens submitted by physicians around the state. Using the soil-sample technique that Waksman so successfully employed, she had by 1948 identified new antifungal agents among the bacteria called actinomycetes. The talents and training of a chemist were needed at this point in the research to look for particular substances possessing antifungal properties. Gilbert Dalldorf, the new head of the division, chose one of his Albany chemists, Rachel Fuller Brown, for this job.

Rachel Brown was born in Springfield, Massachusetts, in 1898. While she was in grade school, her family moved to Webster Groves, Missouri. When she was twelve, her parents separated, and her mother moved back to Springfield, taking Rachel and her brother. When it came time for college, Brown enrolled at nearby Mount Holyoke College thanks to a small scholarship from her high school and financial support from a well-to-do Springfield patroness. At Holyoke she majored in history and chemistry. At the recommendation of Emma Perry Carr, chair of the chemistry department and a well-known mentor of women chemists, Rachel took a master's degree in chemistry at the University of Chicago. Afterward she taught at a private school for girls. She returned to the University of Chicago in 1924 to take a Ph.D. in chemistry and bacteriology. With earnings from teaching school and as a teaching assistant in graduate school, Brown was able to repay the educational grants from which she had benefited. It became one of her lifelong objectives to afford young people the kind of

■ Elizabeth Hazen (left) and Rachel Brown many years after their discovery of nystatin. Courtesy Photography Unit, Wadsworth Center, New York State Department of Health.

educational opportunity she had enjoyed because of the benevolence of others.

Before she defended her doctoral dissertation, she took a job at the Albany headquarters of the New York Department of Health's Division of Laboratories and Research. Several years later she completed that last step for her Ph.D. when back in Chicago representing the division at a professional meeting. By then she was already respected for her work on distinguishing the polysaccharides characteristic of the various pneumococci that cause pneumonia. Before the era of broad-spectrum antibiotics the most successful treatment for pneumonia was the use of blood serums that carried antitoxins (see Paul Ehrlich, p. 17). But each serum was effective against only one bacterial type of pneumonia, hence the need to distinguish the pneumococci.

The partnership between Hazen in New York City and Brown in Albany worked, thanks to the efficiency of the U.S. Post Office in the 1940s. In her New York City laboratory Hazen cultured organisms found in soil samples and tested their activity in vitro against two fungi, *Candida albicans* and *Cryptococcus neoformans*. If she found such activity in a particular culture, she would mail it to Albany in a mason jar. At her end Brown employed painstaking solvent extractions to isolate the active agent in such a culture. (This was before the days of high-performance liquid chromatography and other separation techniques.) After isolating the active ingredient, Brown would return the sample to Hazen in New York, where it was retested against the two test fungi. If it proved effective, its toxicity was evaluated in animals.

Nearly all the agents that killed the test fungi turned out to be highly toxic in animals. Ironically, of the hundreds of soil samples sent to Hazen and Brown from around the world, the one culture that passed all hurdles was originally found in soil from the garden of Hazen's friends, the Walter B. Nourses, and hence named *Streptomyces noursei*. It contained a substance that they first named fungicidin, a name that, unbeknownst to them at that time, had already been used for another substance. They then renamed the substance nystatin in honor of the New York State Public Health Department.

Their division head, Dalldorf, pressed them into making a public announcement of their discovery at a regional meeting of the National Academy of Sciences. The next steps had to involve securing a patent, producing nystatin in large quantities, and testing the drug in human clinical trials—for which the state had no resources. To secure a patent, Dalldorf turned to the Research Corporation, which in those days performed legal services for university and public-sector scientists. E. R. Squibb bought the rights to the patent, conducted clinical trials, and licensed the production and marketing to a wide variety of drug companies. Royalties from these activities were funneled back into the scientific world by the Research Corporation via the Brown-Hazen Research Fund, which gave grants to scientists in the life sciences during the life of the patent.

# 7. Steroids

In the 1930s chemists recognized the structural similarity of a large group of natural substances, including cholesterol, bile acids, sex hormones, the cortical hormones of the adrenal glands, and certain plant substances—the steroids.

The medicinal potential of various steroids quickly became known, but extracting sufficient quantities of these substances, which exist in minute amounts in animal tissue and fluids, was in most cases prohibitively expensive. It was the chemist's ability to create the substances synthetically or semisynthetically, or to modify them in a productive manner—all from comparatively less expensive natural substances—that brought these therapies to large segments of the population.

When the benefits of cortisone to the many people suffering from rheumatoid arthritis became known, it triggered a race to synthesize this substance that was as miraculous in its own way as insulin had been for diabetics. Shortly thereafter a steroid was synthesized that had the same effect on the female reproductive system as progesterone, a hormone that during pregnancy maintains the proper uterine environment and inhibits further ovulation; and so began the first birth-control pill, which produced incalculable changes in our society.

Also indebted to this era of pharmaceutical research is the hormone replacement therapy that has been commonly prescribed for postmenopausal women to prevent osteoporosis and heart problems, although questions have recently been raised regarding its effects. Though not synthesized, the commonly prescribed Premarin, which supplies estrogen to women whose bodies have ceased manufacturing this hormone, is extracted from the urine of pregnant mares by methods dependent on chemistry. In 1998 more prescriptions were written in the United States for Premarin than for any other drug for any purpose.

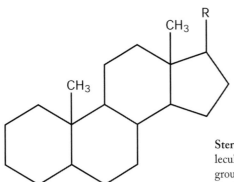

**Steroid nucleus.** All steroids have this general molecular structure in common. Each steroid has a group of atoms that is specific to it, represented by the "R" in the diagram.

# Edward Calvin Kendall (1886–1972), Philip Showalter Hench (1896–1965), Tadeus Reichstein (1897–1996), Max Tishler (1906–1989), and Lewis Hasting Sarett (1917– )

On 20 April 1949 Philip Hench and Edward Kendall announced to a full staff meeting of their colleagues at the Mayo Clinic in Rochester, Minnesota, that they had found among the hormones produced by the exterior of the adrenal gland a substance that would alleviate the most painful symptoms of rheumatoid arthritis, a chronic, crippling disease in which inflammation of the connective tissues causes irreversible damage to the cartilage of the joints. As proof of their claim the two scientists showed dramatic films of twenty-three patients before and after treatment with the substance that the two scientists would soon name "cortisone." These films of patients with severe mobility problems who could again walk and with terrible crippling of their hands who could again write and paint proved equally convincing to their professional peers throughout the world and to the popular press and its readers.

Edward Kendall was born in 1886 in South Norwalk, Connecticut, the only son of a dentist and his wife. As a child Kendall's interest in science led him to build electromagnets and telegraph instruments in his attic workshop. Convinced by his brother-in-law to attend Columbia University, Kendall studied chemistry, a subject to which he had been introduced in high school. In the fall of 1908, after receiving his undergraduate degree, Kendall began his graduate studies, also at Columbia, working with H. C. Sherman on enzymes.

Upon completing his graduate studies, Kendall took a job with Parke, Davis and Company in Detroit. His task there was to isolate the hormone of the thyroid gland. Though the project interested him, he disliked working for a large company and within a few months returned to New York City. In February 1911 he accepted a position at St. Luke's Hospital, where he continued his work on the thyroid. Three years later Kendall left St. Luke's and joined the Mayo Clinic in Rochester, Minnesota, as a biochemist. The Mayo Clinic began in the late 1880s as a group practice that integrated medical research with clinical methods to provide the best possible care for patients.

At the Mayo Clinic, Kendall continued his investigations of the thyroid gland, which produces a hormone that was later named "thyroxine." He succeeded in isolating crystalline thyroxine in 1914 and later attempted to synthesize it. Although Kendall was unsuccessful, this feat was accomplished in 1926 by C. R. Harrington of University College, London.

Kendall then turned to the study of the hormones of the adrenal cortex, the exterior of the adrenal gland. It was known that these hormones could treat adrenal dysfunctions, for example, Addison's disease, an anemic emaciated condition characterized by peculiar brown skin coloring and evidently related to a diseased adrenal cortex. Still these hormones were not fully understood. Using the laboratory equipment of Albert Szent-Györgyi (p. 30), who had spent eight months during 1929–30 working on vitamin C at the Mayo Clinic, Kendall focused on cortin, which was considered to be the main hormone of the adrenal cortex. In the early 1930s Kendall and his colleagues succeeded in isolating a small amount of crystalline material, and soon thereafter made an arrangement with Parke, Davis, which supplied them free of charge with the necessary adrenal glands (the source of cortin) from cattle slaughterhouses. The amount of cortical hormones obtainable was extremely small: less than 250 milligrams from 1,000 pounds of adrenal glands. In return for this supply of starting material the Mayo Clinic agreed to provide Parke, Davis with Adrenalin. This hormone, produced in the interior of the adrenal glands and sold by Parke, Davis (see Jokichi Takamine, p. 47), was easily separated from these glands without affecting the cortin.

By 1940 Kendall and his fellow researchers had determined that no single hormone of the adrenal cortex controls all of its necessary functions. They now deduced that the adrenal cortex produces a number of substances with different functions.

In the meantime Philip Hench, the head of Mayo's Department of Rheumatic Diseases, had been searching for ways to help patients with rheumatoid arthritis. Hench, who was born in Pittsburgh, Pennsylvania, in 1896, was a graduate of Lafayette College and the University of Pittsburgh. He arrived at the Mayo Clinic in 1923, just a few years after completing his M.D. Hench observed that jaundice—the yellowing of the skin and eyes resulting from malfunction of the liver—sometimes alleviated the symptoms of rheumatoid arthritis, as did pregnancy. Hench believed, therefore, that arthritis was reversible, and he attempted

■ Edward Kendall (left) and Philip Hench in a Mayo Foundation laboratory. Courtesy Mayo Foundation.

unsuccessfully to identify the substance that was common to jaundice and pregnancy. He later determined that arthritis sometimes improved after surgery or anesthesia, both of which stimulate the secretion of hormones from the adrenal gland. Thus he began to suspect that the substance he sought originated in the adrenal gland. After discussing the possibilities with Kendall, in 1941 he decided to administer compound E (17-hydroxy-11-dehydrocorticosterone), the most promising of the adrenal hormones, to patients suffering from rheumatoid arthritis. However, it would take many years before this decision could be carried out.

In 1941 the National Research Council (NRC) decided that compound E was important to war work. There were rumors that the Germans already understood the operations of the adrenal cortex and were administering a cortical steroid to Luftwaffe pilots that enabled them to fly at unusually high altitudes with no adverse effects. As in the penicillin program, the NRC formed a committee made up of representatives of research groups at universities, pharmaceutical companies, and other centers of steroid research to organize work and share results. Since extracting the cortical hormones from the adrenal glands of cattle yielded little product for a lot of chemical work and was therefore very expensive, one of the program's major objectives was developing a synthetic—and, it was hoped, cheaper—means of producing the hormones. Work began on synthesizing the less complex compound A, with plans to eventually progress to compound E.

Kendall, who was among those recruited for the work, started with desoxycholic acid (Figure 1), which was available from cattle bile in relatively plentiful quantities. He attempted by a series of chemical reactions to move the oxygen atom on one of the carbons (C-12) in the desoxycholic acid molecule to its neighbor (C-11). In January 1942 Merck and Company, which was working in close cooperation with the Mayo Clinic, sent the newly hired Lewis Sarett to confirm Kendall's results. Sarett, a graduate of Northwestern University, where his father was a professor of speech, had just completed his Ph.D. at Princeton in a mere two-and-one-half years. He was already well acquainted with steroid research since he had written his doctoral dissertation on how to effect reasonably inexpensive syntheses of

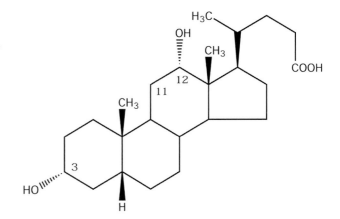

**Figure 1.** Desoxycholic acid.

female hormones. The generation gap between Sarett and Kendall was bridged to some extent by their common passion for chess. But after three months in Kendall's laboratory and even more time back at Merck in Rahway, New Jersey, Sarett was getting nowhere with proving by chemical means that Kendall's work had indeed moved the oxygen to C-11 (this was before the introduction of infrared spectroscopic techniques). Spurred by an ominous remark of his boss's—"Well, maybe Merck isn't the company for you"—Sarett then set about trying to prove the negative, that there was *no* carbonyl (C=O) at C-11—a difficult decision since Kendall was such a famous chemist and so certain of his results. With this new tactic Sarett was able to show that molecular changes had occurred, but not the desired one.

In 1943 the Swiss chemist Tadeus Reichstein, who was also part of the consortium seeking to understand and synthesize cortical steroids, succeeded in preparing compound A (Figure 2). Reichstein was born in Poland in 1897, the son of an engineer. After his family moved to Zurich, Reichstein became a naturalized citizen of Switzerland. He studied chemistry at the Eidgenössische Technische Hochschule, completing his degree in 1920. He returned a year later to pursue graduate studies with the famous polymer chemist and future Nobel laureate Hermann Staudinger.

Reichstein spent the first years of his scientific life studying the aromatic substances in roasted coffee, but in 1931 he

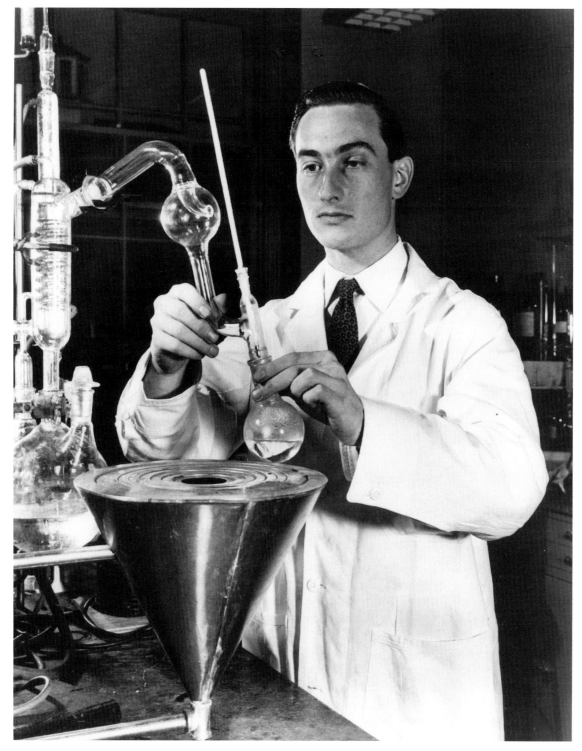

■ Louis Sarett with a distillation apparatus, in 1950. Courtesy Merck Archives, Merck & Co., Inc.

left this work to become an assistant to another future Nobel laureate, Leopold Ruzicka. In 1933 Reichstein succeeded in synthesizing vitamin C, independently of the British chemist Sir Norman Haworth, who received the 1937 Nobel Prize in chemistry for his work on that vitamin. A year later Reichstein accepted a post at the University of Basel.

Reichstein, too, was fascinated by the adrenal gland, and by 1936 he had identified four different compounds from the adrenal cortex. Later, many more compounds were identified (twenty-nine in all), but researchers were only interested in the six believed to be biologically active. In 1936 Reichstein demonstrated the similarities of these hormones to androgens (male sex hormones), thereby providing evidence that the adrenal compounds, too, were steroids.

In 1944 the NRC committee on the adrenal gland disbanded because the yield of compound A that Reichstein had achieved was so low (0.04%, that is, less than one tenth of a percent) compared with the amount of starting material; the committee concluded that any larger-scale synthesis of it or of compound E would be prohibitively expensive. Only the Merck and Mayo groups continued working with adrenal hormones. Two other Merck chemists, Jacob van de Kamp and Stewart Miller, from Max Tishler's development group, were dispatched to Rochester with the charge of producing sufficient compound A so that clinical trials could go forward. This synthesis actually incorporated some critical steps from Kendall's failure, and it produced far greater yields than Reichstein's synthesis. But by 1946 clinical trials showed that compound A was clinically ineffective.

Meanwhile Kendall often visited Sarett when he was in Rahway for research meetings, but his particular purpose in Sarett's laboratory was to drop off the next move in their ongoing chess games. They were also participants in another game, however: in 1944 Sarett managed to convert compound A to compound E. For this conversion to take place, a hydroxyl group (-OH) had to be added to C-17. Rather like a chess game, there was no direct way of making this happen; in fact, the whole synthesis took thirty-seven steps. The yield was too small for the process to be commercially viable, less than Reichstein's yields of compound A, but Sarett's synthesis of cortisone (Figure 3) did confirm the structure of the compound.

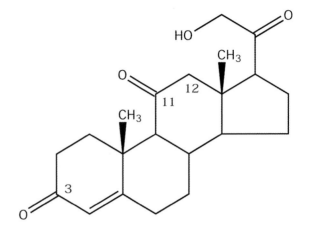

**Figure 2.** Compound A—11-dehydrocorticosterone.

With only this difficult synthesis in hand, the Mayo and Merck groups decided to develop large-scale syntheses of compound E so that clinical trials could proceed—Merck being expected to produce the lion's share of the product. Max Tishler, then Merck's director of development research, was key in this effort.

Tishler was born in Boston, the fifth of six children of parents who had immigrated from Germany and Romania. Because his father abandoned the family, Tishler grew up in economically straitened circumstances. He contributed to the family's finances from childhood on. As a delivery boy

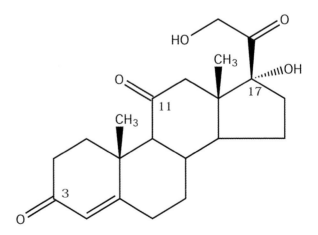

**Figure 3.** Compound E—17-hydroxy-11-dehydrocorticosterone. Cortisone.

■ Tadeus Reichstein, in May 1959. Copyright Kurt Wyss Diplomierter Photograph.

for a pharmacy at the age of twelve, he brought medicines to the sick and dying during the "Spanish" influenza epidemic that killed over half a million people in the United States in 1918—an unforgettable experience that shaped his career decisions. During his years as a scholarship student at Tufts University and then as a graduate student in the chemistry department at Harvard University, he continued to work in pharmacies, even passing the examination to become a certified pharmacist. He finished his Ph.D. in the middle of the Depression and was kept on in various junior teaching appointments, but he was delighted when his adviser arranged an appointment for him in 1937 at the Merck Research Laboratory.

His first important job at Merck was to design a new synthesis for vitamin $B_2$, riboflavin, that would not infringe existing patents. Merck wanted to build on its success with vitamin $B_1$ (see Robert R. Williams, p. 33) by moving into production of $B_2$, but IG Farben and Hoffmann–La Roche, the firms that held patents for synthesizing it, would not license Merck to produce it. Tishler personally supervised the scaling up of his synthesis—a job normally handed over to development specialists. But Tishler experienced the joy of being present when the first few drops of vitamin $B_2$ came out of the $5 million plant.

His talents in process development were recognized by a series of promotions. During World War II, Tishler directed work on several processes, including those for producing sulfa drugs, streptomycin, penicillin, and cortisone. One of his tactics was to divide his staff into teams, each focusing their energies on a certain group of steps in the process being improved. For example, in the cortisone synthesis, he did not want to continue to use osmium tetraoxide to introduce the hydroxyl (-OH) group on C-17 since it was an expensive and dangerous reagent. The team was eventually able to establish the conditions for using commonly available permanganate instead. In the end the cortisone synthesis took some twenty-six steps (reduced from thirty-seven). This process, with a starting material of cattle bile, continued to produce competitively priced cortisone even after the conclusion of the great race to synthesize the hormone from other starting materials (see Percy Julian et al., p. 106, and Robert Burns Woodward, p. 114).

In 1948 Kendall and Hench—who had recently returned to the Mayo Clinic after serving in the Medical Corps during World War II—decided to use compound E to treat one patient suffering from rheumatoid arthritis who had become despondent with her lack of improvement after the usual treatments. The results were astounding, and they expanded their trial with Merck-supplied hormones; all of the patients responded favorably to compound E. The board of governors of the Mayo Clinic decided to conduct one final trial, for which they appointed a committee. They first administered a placebo of cholesterol, noted the patient's response, and then administered compound E. This test unequivocally demonstrated that compound E had clinical potential. In 1949 the clinical results were announced, and Kendall and Hench named compound E "cortisone" as a contraction of its chemical name.

In 1950 Kendall, Hench, and Reichstein were awarded the Nobel Prize in physiology or medicine "for their discoveries relating to the hormones of the adrenal cortex, their structure and biological effects." Although cortisone is not a cure for rheumatoid arthritis, it does relieve the effects of the disease and has also been found effective in treating Addison's disease (caused by the failure of the adrenal gland to create cortisone), asthma, and allergy-related skin conditions. Prolonged use of cortisone, however, does bring side effects, particularly symptoms of Cushing's disease, an ailment caused by overactivity of the adrenal gland and most noticeable by its symptom of bloating. Because of these serious side effects, pharmaceutical companies relied more on hydrocortisone, another cortical hormone, and its modifications and developed synthetic analogs of cortisone and nonsteroidal anti-inflammatories.

Shortly after receiving the Nobel Prize, Kendall left the Mayo Clinic (having reached the mandatory retirement age) and became a visiting professor in the chemistry department of Princeton University. There he continued his work on the adrenal cortex, focusing primarily on its nonsteroidal compounds, until his death in 1972.

Hench remained at the Mayo Clinic until 1957, although his participation in clinical testing declined significantly. Much of his time was spent writing articles and reviews pertaining to his Nobel Prize–winning work. Hench died in 1965.

Reichstein continued his investigations, begun in the late

■ Max Tishler performing a filtration, in 1938. Courtesy Merck Archives, Merck & Co., Inc.

1940s, on toxic substances in plants, insects, and toads, examining their relationship to steroids. In 1960 he became the director of the Institute of Organic Chemistry at the University of Basel. Although he became director emeritus seven years later, Reichstein continued to be actively involved in research until shortly before his death at the age of ninety-nine in 1996.

After the triumph of cortisone Tishler became president of Merck, Sharp and Dohme Research Laboratories in 1957 and senior vice president of research and development at Merck and Company in 1969. At Merck he directed the development of important drugs to control high blood pressure; vaccines against measles, mumps, and rubella (German measles); and important antibacterial and antiparasitic agents. When Tishler retired from Merck, he took an appointment at Wesleyan University in Middletown, Connecticut.

Immediately after his partial synthesis of cortisone Sarett joined in the competition to develop a shorter, cheaper synthesis of cortisone. He succeeded by creating the first total synthesis of the hormone from simple starting materials. Later, when the undesirable side effects of cortisone were noticed, he devoted his research talents to developing steroidal analogs of cortisone and nonsteroidal anti-inflammatories. He rose to be president of Merck, Sharpe and Dohme Research Laboratories, achieving this objective in 1969, and then became Merck's senior vice president for science and technology. After his retirement he remained active in directing and consulting for biotechnology companies.

## PERCY JULIAN (1899–1975), RUSSELL MARKER (1902–1995), AND CARL DJERASSI (1923– )

Relying on expensive animal sources for hormones was rapidly perceived as a bottleneck in the push to discover further medicinal uses of hormones and to provide therapeutic quantities of these substances to a significant number of patients at a reasonable cost.

Percy Julian, Russell Marker, and Carl Djerassi were among those scientists who participated actively in the synthesis and large-scale production of natural steroids—both cortisone and the sex hormones—from plant compounds (see also Robert Burns Woodward, p. 114).

Percy Julian was born in 1899 in Montgomery, Alabama—the son of a railway mail clerk and the grandson of slaves. In an era when African Americans faced prejudice in virtually all aspects of life—not least in the scientific world—he succeeded against the odds. Inadequately prepared by his high school, he was accepted at DePauw University in Greencastle, Indiana, as a sub-freshman—meaning that he had to take high school courses concurrently with his freshman courses. He majored in chemistry at DePauw, graduating as valedictorian of his class. After graduating, he worked at Fisk University for two years as a chemistry instructor. He then won an Austin Fellowship to Harvard University and completed a master's degree in organic chemistry there. He returned to teaching at West Virginia State College and Howard University. In 1929 Julian began his studies on the chemistry of medicinal plants at the University of Vienna, an endeavor funded by a grant from the Rockefeller Foundation. In 1931, with doctoral degrees in hand, he and a Viennese friend,

Josef Pikl (whom Julian had invited and helped come to the United States), took positions back at Howard and two years later moved to Julian's alma mater, DePauw. There they accomplished the first total synthesis of the active principle of the Calabar bean, physostigmine (or eserine), used since the end of the nineteenth century to treat glaucoma. Physostigmine, an alkaloid, eases the constriction of outflow channels from the eye's aqueous humor to relieve high pressure there—which, if left untreated, damages the retina and eventually causes blindness.

Meanwhile researchers in many countries were looking for ways to synthesize steroids, including cortisone and the sex hormones. Around this time German chemists discovered that the steroid stigmasterol, which Julian had obtained as a by-product of the physostigmine synthesis but was also obtainable from soybeans, could be used in the preparation of certain sex hormones. In pursuit of this lead, in 1936 Julian wrote to the Glidden Company in Chicago, requesting samples of soybean oil. By a curious quirk of fate he wound up working for Glidden instead. Glidden's vice president, W. J. O'Brien, was also on the board of the Institute of Paper Chemistry in Appleton, Wisconsin, where Julian had recently applied for a job as a research chemist. O'Brien listened to his fellow board members bemoan the difficulty in hiring Julian because of a law forbidding Negroes to stay overnight in Appleton. "If he is half as good as they say he is," said O'Brien, "I can use him at Glidden." Julian was promptly made director of research of the Glidden Soya Products Division, where he remained until 1954, when he founded his own company, Julian Laboratories of Franklin Park, Illinois, and Mexico City (which he eventually sold to Smith, Kline and French).

Three years after he came to Glidden, plant workers reported that water had leaked into a tank of purified soybean oil and formed a solid white mass. Julian immediately identified the substance as stigmasterol and designed an innovative industrial process that produced five to six pounds per day of progesterone (worth thousands of dollars in those days). Soon other sex hormones were in production.

When in 1948 Hench and Kendall at the Mayo Clinic announced that their compound E (cortisone) had such remarkable effects on rheumatoid arthritis, Julian jumped into the exciting competition to synthesize cortisone

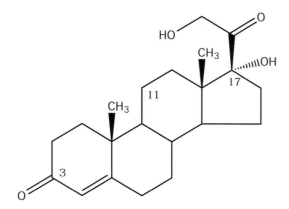

**Figure.** Substance S, synthesized by Percy Julian, among others. It differs from cortisone only in that it lacks an oxygen atom on C-11.

■ Percy Julian, at Julian Laboratories, Inc. Courtesy DePauw University Archives and Special Collections.

inexpensively. The result was a new synthesis for Reichstein's Substance S (Figure), which is also present in the adrenal cortex and differs from cortisone only in that it lacks an oxygen atom at C-11 (see Edward Kendall et al., p. 97). From this substance he synthesized both cortisone and hydrocortisone. Hydrocortisone and its derivatives today are more widely prescribed than cortisone products, and most industrial syntheses still begin along the same route that Julian pioneered.

Julian placed a high priority on his contributions to society as a scientist and a citizen. He was particularly active in groups seeking to advance conditions for African Americans, helping to found the Legal Defense and Educational Fund of Chicago as well as serving on the boards of several other organizations and universities. He was always attempting to build bridges between diverse groups of people.

Like Julian, Russell Marker turned to plant substances from which to synthesize steroidal hormones.

In the great American tradition of rugged individualism, Russell Marker was born in 1902 in a log cabin on a farm near Hagerstown, Maryland, where his father was a sharecropper until he could afford to purchase his own land. Although Marker's father wanted him to stay on the farm, his mother encouraged her son to pursue a college education. In this she prevailed. Since the high school that Marker attended offered only a commercial program, he had not taken the preparatory subjects necessary for admission to the University of Maryland, but he did well in summer-school mathematics and English courses and was accepted to the regular degree program in the fall of 1919. His introduction to the subject of chemistry, unlike that of most other students, came in college, where he had to catch up with his classmates' familiarity with simple equipment like test tubes and beakers and with the chemical symbols. In the organic chemistry laboratory he demonstrated outstanding skill in carrying out difficult organic syntheses, a talent that would serve him well throughout his career. On completing his bachelor's degree, he continued at the university as a graduate student, writing both a master's thesis and a doctoral dissertation. When he learned that he lacked required courses in physical chemistry, he left without completing the degree. His interest was in organic chemistry, and he thought he knew all the physical chemistry he needed.

Setting out in industry, Marker worked for about half a year as an analytical chemist at the Naval Powder Factory in Indian Head, Maryland. In 1926 he obtained a more research-oriented position at the research laboratory of the Ethyl Gasoline Corporation in Yonkers, New York. This company had recently been founded to produce tetraethyl lead, the gasoline additive that increased the fuel efficiency of gasoline engines and made usable the petroleum fractions previously burned or vented into the atmosphere. (Fifty years later the Environmental Protection Agency instituted the reduction of tetraethyl lead in gasoline as hazardous to the health of gasoline attendants and others who came into close contact with its fumes.) At Ethyl, Marker helped develop the standard gasoline against which "octane" is rated. Meanwhile word spread of his reputation as a wizard at synthesizing organic chemicals that other scientists found difficult to obtain, and he was asked to synthesize a particular compound for a scientist working at the Rockefeller Institute in New York City. After Marker successfully completed this task, Simon Flexner, the institute's president, invited him to take a position at this prestigious biomedical research facility in 1928.

Although Marker's major responsibility at the institute was to synthesize compounds for other scientists, he was also encouraged to carry out his own research projects. In the early 1930s Marker became interested in the emerging field of hormones. When he declared his interest in synthesizing hormones from plant materials, he came into conflict with the powers at the institute. One of the institute scientists, working on the plant steroid present in the sarsaparilla root (the root from which "root beer" flavor is derived), had supposedly "proved" that the molecule simply could not be converted into a chemical intermediate to a hormone. In a confrontation with Flexner, who praised Marker's performance in general but would not agree to a research project that looked like it was going nowhere, Marker threatened to leave if he could not have his way.

And leave he did in 1934—for a much lower-paying position in the chemistry department at Pennsylvania State University. Here he was welcomed by department chairman Frank Whitmore, who had known and admired Marker's work on hydrocarbons at Ethyl. With the aid of reagents and intermediates supplied by Parke, Davis and Company, Marker

■ Russell Marker in his university laboratory, in 1940. Courtesy Penn State University Archives, Eberly Family Special Collections Library.

set to work repeating the experiments done on steroids by the great German and Swiss scientists like Adolf Butenandt and Leopold Ruzicka. Marker's first big success was in isolating pregnanediol from bull's urine. From this substance he synthesized progesterone, the hormone that, among other important functions, naturally inhibits the ability of a pregnant woman to become pregnant again during gestation.

Meanwhile Marker was able to show that the structure of one of the sapogenins, plant sterols modified by treat-ment with acid, was not that reported by leading chemists in Germany and the United States—and that the compound was not inert. (The word *sapogenin* derives from the property of these substances to foam in water, like soap.) He then proceeded to transform the sapogenin from sassafras root, sarsasapogenin, into a molecule identical in structure to that of progesterone. Marker turned next to a substance called diosgenin, obtained from some Japanese scientists who had extracted it from a member of *Dioscorea*, a genus containing the yams, and transformed it into progesterone. This success sent Marker off on a wide-ranging search for species of this plant group with sufficiently large roots to make harvesting and processing the material economical. In 1942 he searched for and found in the wilds of Mexico the *cabeza de negro*, which has roots that can weigh up to a hundred pounds each. At the time Mexico seemed about to join the Axis powers in World War II, which made his adventure appear risky indeed to officials at the United States embassy in Mexico City.

Back in the United States with samples of dried root, Marker tried unsuccessfully to convince the president of Parke, Davis, Alexander Lescohier, that the company should rely on diosgenin as the starting material for their manufacture of hormones. Lescohier was prejudiced against Mexico because he had undergone a botched appendectomy in that country years before, so he turned Marker down. He would rather depend on bull's urine and sassafras roots for his supply! Drawing half his meager savings from the bank, Marker returned to Mexico to pursue the development of diosgenin on his own. He made arrangements for the harvesting, drying, and extracting of a syrup from the roots and its export to the United States to a small laboratory, which turned it into what amounted to three kilograms of progesterone—the largest amount of progesterone that anyone had ever seen up to that time.

While he was in Mexico, Marker tried to interest financial backers in his venture—without success, until he found in the Mexico City phone book an entry for Laboratorios Hormona. This company, begun in 1933 to extract hormones from animal sources, was founded by Emeric Somlo, a Hungarian lawyer-entrepreneur who had first come to Mexico in the 1920s with a Hungarian pharmaceutical company. In March 1944, Marker, Somlo, and Frederico Lehmann, Hormona's chief chemist and also a Hungarian, formed a new company called Syntex (the "ex" being included to give the corporate name a Mexican sound).

Somlo and Marker did not get along well, and in May 1945 Marker left the company, taking his processes with him. Fortuitously, Somlo was able to recruit George Rosenkranz, a fellow Hungarian refugee and an organic chemist of great ingenuity, who managed to get the factory up and running within two months of his arrival with processes of his own invention. Meanwhile, Marker founded Botanica-Mex, which also made hormones from *Dioscorea*, principally *D. barbasco,* and did so with some success. But in 1949 Marker left Botanica-Mex, which went on to become Diosynth, a subsidiary of SmithKline and French, then of Organon, and most recently of Akzo-Nobel.

In 1949, at the age of forty-seven, Marker left chemistry altogether to set up a business making reproductions of eighteenth-century silver objects. He had become a legend in his own time. He first reappeared in the chemical community in 1969 when the Mexican Chemical Society presented him with an award. Some at the ceremony presumed that this was a posthumous award and were shocked to see Marker walk in. In fact he lived another two-and-a-half decades, occasionally attending symposia honoring him and his chemical colleagues.

Carl Djerassi was one of Julian's many rivals in the competition to synthesize cortisone, and he eventually followed Marker at Syntex. Djerassi came to the United States in 1939, after fleeing the Nazis in Austria. He sped through Tarkio College in Tarkio, Missouri, after which he spent a brief period working on antihistamines for the Swiss pharmaceutical company CIBA at its New Jersey facility. He then completed a doctorate in organic chemistry at the University of Wisconsin, where he wrote his dissertation on how to transform the male sex hormone testosterone into the female sex hormone estradiol, using a series of chemical reactions. Djerassi's longtime fascination with steroids prompted his return to CIBA, but he was not allowed to work on steroid synthesis there. That promising field of research was reserved for the laboratories at CIBA's corporate headquarters in Switzerland. Djerassi was disappointed, and in 1949—shortly after Julian's paper on Substance S appeared—he joined Laboratorios Syntex S.A. in Mexico City.

Syntex was now trying to synthesize cortisone from diosgenin, a steroidal substance found in Mexican yams; the company had already produced male and female sex hormones from diosgenin. In 1951 Djerassi's group successfully synthesized cortisone, improving on a procedure originally developed in 1944 by Lewis Sarett (p. 97) of Merck: the Syntex process not only used a cheaper raw ma-

■ Carl Djerassi at Syntex in 1951, with his assistant
Arelina Gonzalez. Courtesy Professor Carl Djerassi.

terial but also required only about half as many steps. Soon
other groups reported their syntheses. In the end Upjohn
was most successful commercially, with its use of a microorganism to convert progesterone to cortisone, but Syntex
benefited, too, because it was commissioned to supply the
progesterone to Upjohn.

In 1951, the same year that Djerassi's group at Syntex
synthesized cortisone, it also synthesized the first effective

oral contraceptive. It had long been known that during
pregnancy progesterone serves as a natural contraceptive by
inhibiting further ovulation while maintaining the proper
uterine conditions for the fetus. But taking natural progesterone orally weakens its biological activity, and thus the
search was on for a more active sex hormone that could survive digestive processes. A compound synthesized at Syntex
proved to be one of the most potent oral progestins ever

made. The U.S. Food and Drug Administration approved Syntex's norethindrone as well as a related drug synthesized by Frank B. Colton at G. D. Searle and Company, first as a treatment for menstrual difficulties (1957) and then as an oral contraceptive (1960). Colton's norethynodrel differed from norethindrone in the location of just one double bond in the molecule.

Djerassi maintained a twenty-year-long relationship with Syntex, while also accepting academic appointments after his 1951 triumphs, first at Wayne State University in Detroit and then at Stanford University. He made many more advances in synthetic organic chemistry and refined the techniques of mass spectroscopy and methods for deducing the precise orientation in space of the atoms in a molecule from optical rotatory dispersion.

Like Julian, Djerassi is a social activist as well as a scientist. In line with his work on the birth-control pill, he seeks to raise consciousness about the global need for population control. Stemming from his years with Syntex have been his efforts to encourage science in developing countries like Mexico. In memory of his daughter, who was an artist, and consonant with his own artistic and literary interests, he has established a colony for artists near Santa Cruz, California.

# 8. Drugs Affecting the Central Nervous System: Antipsychotics and Tranquilizers

The treatment of mental illness has changed significantly over time. In the early twentieth century the key innovation was the introduction of psychotherapy. Psychological ailments had long been treated with drugs, but this method produced highly problematic outcomes.

By mid-century, for example, patients with relatively mild neuroses as well as those with serious psychoses were commonly given barbiturates or amphetamines, both highly addictive classes of substances, to sedate or stimulate them as their illnesses seemed to require. Institutionalization was inevitable for thousands of patients. Any kind of communication, therapeutic or otherwise, with the sickest patients remained virtually impossible. To control the destructive, often suicidal, conduct of patients, the staff members of mental institutions commonly had to resort to such extreme measures as straitjackets, electric shock treatments, and prefrontal lobotomies.

In the 1950s the first really effective drugs for treating mental illness came on the scene. Among these pioneer drugs were the antipsychotics reserpine and Thorazine (chlorpromazine), used for schizophrenia, and the tranquilizers Librium (chlordiazepoxide) and Valium (diazepam). In the same decade other important drugs were discovered, including lithium for treating manic-depressive illness and Tofranil (imipramine) to lift the burden of depression. Several of these drugs were discovered almost accidentally. But they were soon used as probes to develop a more intimate understanding of the biochemistry of the central nervous system so

that the following generations of drugs for treating mental illness—for example, Prozac (fluoxetine) (see Ray W. Fuller et al., p. 167)—could be more rationally designed.

The introduction of these medicines and the consequent reduction of the most destructive symptoms of mental illness had immense social consequences. Those patients who had to remain hospitalized were now able to communicate better and so could benefit from psychotherapy. Thousands of former inpatients could now be treated in their home communities, and some never needed to enter hospitals, so long, that is, as they followed special medication regimens. But emptying out mental hospitals had some negative consequences as well, because an adequate number of community outpatient clinics were not constructed to treat such patients and, even if they had been, the nature of the patients' diseases made them unlikely to take medication on schedule or keep regular appointments. Another important social consequence of the introduction of these new medications was the growing acceptance of the view that mental illness is as much a biochemical phenomenon (nature) as it is a social and cultural one (nurture) and that indeed these two forces can and do interact.

# Robert Burns Woodward (1917–1979)

Reserpine, introduced in the United States in 1953, and chlorpromazine, introduced in 1954, initiated the psychiatric revolution that continues to evolve to this day. The path to reserpine was quite different from the route to chlorpromazine, which eventually became the more frequently prescribed medicine because it had fewer undesirable side effects. Reserpine came out of the tradition that stretches back to the early nineteenth century of isolating active principles from natural materials and the much more recent capability of chemists to synthesize complex naturally occurring molecules. Chlorpromazine, by contrast, descended from the tradition of making new medicines by modifying synthetic dyestuffs to make parts of their molecules conform to structures known to have the activities desired in the new medicine—an approach that by the 1950s had become much more systematic than in earlier days.

*Rauwolfia serpentina,* or snake root, had long been used in Hindu medicine for a wide variety of purposes. Beginning in the 1930s, Indian and western European chemists worked to isolate the most active substance in *Rauwolfia*'s roots—an alkaloid named reserpine. Reserpine was found to lower blood pressure and calm agitated psychiatric patients while permitting them to be aware of their surroundings and to carry on a relatively normal level of mental activity compared with the total lethargy brought on by other sedatives like the barbiturates. As in many other extracts from plant or animal sources, attempts were soon made to synthesize reserpine to bring down costs and to ensure the purity and uniformity of the drug. The research became more urgent when in the spring of 1955 the Indian government embargoed the export of snake root in response to increased demand.

In 1956 the organic chemist Robert Burns Woodward, at Harvard University, succeeded in synthesizing this molecule, the most complicated synthesis up to that time (see Figure). Reserpine was one among a succession of analyses and syntheses of increasingly complicated molecules, including chlorophyll, that won Woodward the Nobel Prize in chemistry in 1965.

Woodward was born in Boston, Massachusetts, in 1917. His father died the following year, and although his mother remarried, she was soon abandoned by her second husband and left to bring up her son in straitened circumstances. When he was eight years old, he received a chemistry set, which he improved by adding more reagents and gathering other instructions, including those gleaned from a college-level textbook.

When he enrolled at the Massachusetts Institute of Technology in 1933 at the age of sixteen, Woodward knew as much organic chemistry as a college senior. After a promising freshman year he was given his own small laboratory. Then he almost forfeited his opportunities by skipping chemistry classes and neglecting other subjects. At the end of the first semester of his sophomore year he was asked to withdraw from college. He spent a semester as an employee

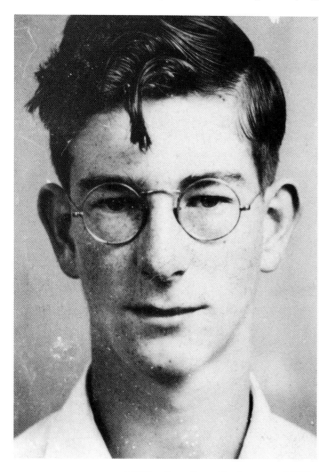

■ Woodward as a freshman at MIT. Courtesy MIT Museum.

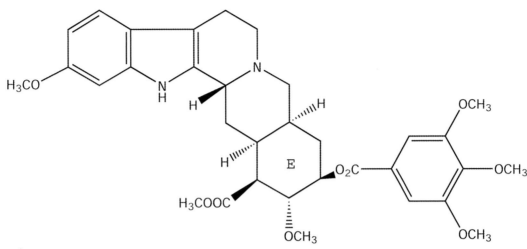

**Figure.** Reserpine.

of the MIT biology department and was then permitted to re-enroll. In four years Woodward completed both his bachelor's degree and his Ph.D.—academic work that usually takes at least seven years.

After teaching summer school at the University of Illinois, he returned to Harvard to take a position as a research assistant in the chemistry department. He remained at the university for the rest of his career, progressing from this lowly position through the usual academic ranks to two prestigious endowed chairs. While he was a faculty member, he also served as a consultant to a number of companies, most enduringly the pharmaceutical companies Pfizer and Company and CIBA-Geigy.

During World War II, Woodward participated in war-related research on two drugs that were in short supply: quinine and the new "wonder drug" penicillin. Neither of these endeavors resulted in a synthesis that could be scaled up to produce the necessary quantities. But he developed approaches and understandings that in the long run helped him to become the creator of the most famous syntheses ever.

Woodward's work on the synthesis of the antimalarial quinine was supported by the Polaroid Corporation of Cambridge, Massachusetts, although Polaroid was not a pharmaceutical company. Quinine, because it has the property of polarizing light, was the basis for Edwin Land's original business of making light polarizers of all types (hence the name Polaroid). Woodward quickly finished the job for which he was hired—finding new light polarizers to replace

quinine. He then petitioned Land to support his synthesis of quinine. With the assistance of William von Eggers Doering he succeeded in synthesizing what was thought at the time to be the immediate precursor of quinine—a feat enthusiastically applauded in the popular press.

In the Allied effort to produce penicillin, Woodward was engaged in determining the molecular structure of the drug—structural analysis being the usual prerequisite to synthesis. On the basis of information gained from reacting penicillin with various chemical reagents, which resulted in products of known structure, he correctly reasoned that penicillin must include the four-membered beta-lactam ring structure, and he thus weighed in on this side of the debate about penicillin's structure (see Alexander Fleming, p. 73, and Howard Walter Florey and Ernst Boris Chain, Figure 1, p. 81). His hypothesis was confirmed by Dorothy Hodgkin's X-ray crystallographic work (see p. 82).

After the war, among other consultancies, Woodward advised chemists at Pfizer, who were trying to determine the structures of the natural antibiotics Terramycin and Aureomycin; later he was an adviser when they synthesized tetracycline. Merck and Company agreed to support two postdoctoral students to work in Woodward's laboratory on the total synthesis of steroid hormones, with Monsanto, for whom he was consulting, supplying starting materials and intermediates. Woodward's synthesis of cortisone was among those announced in 1951 (see Lewis Hastings Sarett, p. 97, and Percy Julian et al., p. 106, for other contenders). In 1956

115

■ Robert Woodward, with his trademark cigarette, sitting at a desk with a statue of a Scotsman. Courtesy Harvard University Archives.

Woodward synthesized lysergic acid, a derivative of ergot (see Henry Hallett Dale and Otto Loewi, p. 59), while working as a consultant to Lilly Research Laboratories, which was considering LSD (lysergic acid diethylamide) as an aid to psychotherapy. A French fine-chemicals company, Roussel-Uclaf, and a Swiss pharmaceutical company, Sandoz Ltd., used Woodward's synthesis of reserpine, also accomplished in 1956, under a patent administered by the Research Corporation. (For more on the Research Corporation, see Robert R. Williams, p. 33, and Elizabeth Lee Hazen and Rachel Fuller Brown, p. 95).

Woodward's reputation was as a chemist who planned syntheses more systematically than anyone ever had before. Besides incorporating the latest theories on molecular structure and reaction mechanisms and using the most modern analytical instrumentation, he applied an uncanny intuitive sense of what was possible in the molecular world, as if he lived there part of the time.

The reserpine molecule (Figure), for example, includes six ring structures and six chiral centers. (A chiral center is an atom, carbon in this case, that has four different attached chemical groups that must be arranged in a particular relationship to one another in order to mimic the natural molecule.) One of the rings contains five of the compound's chiral centers. Woodward decided to confront that ring first by reacting two simple reagents using a method he had known of since boyhood, the Diels-Alder reaction. The new ring that was produced had the carboxyl group ($-CO_2H$) in the right place for reserpine, a double bond that could be used later for the introduction of oxygen groups, and three of the required chiral centers. He was confident because of his previous experience with the Diels-Alder reaction that the hydrogen atoms on these chiral centers would be oriented correctly.

In the reserpine and other syntheses Woodward directed graduate and postdoctoral students in carrying out the experiments he designed. From the time of his work on quinine Woodward's international reputation attracted chemists from all over the world. In his lifetime he guided over four hundred graduate and postdoctoral students. His habit of staying in his office until all hours and the length of his Thursday-night seminars were legendary. But his sense of humor about his role as leader helped weld the Woodward group of postdoctoral students together and made collabo-

rating with him more like a game. Indeed he thoroughly enjoyed virtually all nonathletic diversions, including poker games, which he called "probability sessions."

The syntheses being carried out in Woodward's own laboratory grew ever more complex. The synthesis of colchicine, the substance derived from the autumn crocus that had long been used to control the symptoms of gout, was one example. He therefore found it difficult to do justice to all his commitments, and about 1960 he bowed out of a number of his consultancies. He was also at that time looking for a less-restricted research environment. In 1963 CIBA Limited, soon to become CIBA-Geigy, established the Woodward Research Institute for him in Basel, Switzerland. The arrangement between CIBA and Woodward brought him to Switzerland for six weeks a year. At other times he was in close contact with his dozen or so researchers through weekly progress reports and telephone calls.

Vitamin $B_{12}$ was Woodward's most complex synthesis. This vitamin, part of the B complex, is essential for the normal functioning of the human body; without it, pernicious anemia develops. The experiments were done partly in the Harvard laboratories and partly under the direction of Albert Eschenmoser at the Eidgenössische Technische Hochschule (ETH) in Zurich, Switzerland. The synthesis, which began in 1960, took over a dozen years and involved about one hundred people from twenty nations—mainly ever-changing populations of postdoctoral students—working in the two research groups.

The $B_{12}$ synthesis was also an outstanding example of the role of natural-products synthesis in generating fundamental knowledge in organic chemistry. Problems in the synthesis led to the formulation of the Woodward-Hoffmann rules, which permit chemists to predict, for example, the products that will form when two compounds are activated by heat compared with the distinct products that form when the same two compounds are activated by light. For his contribution to this work Roald Hoffmann received the 1981 Nobel Prize in chemistry (shared with Kenichi Fukui, a Japanese chemist who developed a different way of analyzing molecular interactions that could explain the same chemical changes). Woodward died before the Nobel award was given; had he been alive, there is little doubt that he would have shared the award.

# PAUL CHARPENTIER, HENRI-MARIE LABORIT (1914–1995), SIMONE COURVOISIER, JEAN DELAY (1907–1987), AND PIERRE DENIKER (1917–1998)

The story of chlorpromazine begins in 1937 when Daniel Bovet (see p. 68) at the Institut Pasteur created the first synthetic antihistamine, which was unfortunately too toxic to be used as a drug to alleviate allergic symptoms. Subsequently, the pharmaceuticals division of the French company Rhône Poulenc (now Aventis) had been cooperating with other scientists at the Institut Pasteur on developing new antihistamines.

One research group at Rhône Poulenc, headed by the chemist Paul Charpentier, was trying to create an improved antihistamine loosely modeled on diphenhydramine, which was already known and is today the active ingredient in such drugs as Benadryl, Dramamine, and Sominex. Charpentier actually began with phenothiazine (Figure 1), a similarly structured molecule that started out life as a synthetic dye and then was used as a worming agent in veterinary practice. The new molecule that he synthesized, promethazine (Figure 2), was commercialized as an antihistamine. But in common with other antihistamines, the drug's side effects seemed to point to activity in the central nervous system; so Charpentier began to focus on enhancing these activities. Meanwhile, samples of promethazine were sent to Henri Laborit, a surgeon in the French Navy who was then stationed in Morocco and who was looking for an agent to add to the cocktail of drugs that he administered to patients before anesthesia.

Laborit was born in Hanoi, then in the French colony of Indochina (now Vietnam), and educated at Carnot University in Paris and at the Naval Health Service in Bordeaux. He had seen service on board ship in the Mediterranean and in North Africa during World War II and after. He was particularly interested in creating a kind of artificial hibernation, which included lowering the body's temperature, to lessen the danger of shock during and after surgical operations. Indeed Laborit reported promising results from promethazine in this area.

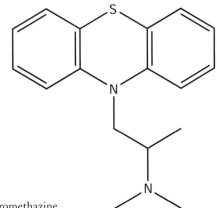

Figure 2. Promethazine.

Among the variations on the molecule that Charpentier tried was one that introduced a chlorine atom into one of the rings of promethazine, thus forming chlorpromazine (Figure 3). In December 1950 he sent a sample of this material to Simone Courvoisier, the head of pharmacology at Rhône Poulenc. In one of the tests, rats that had been conditioned to climb a rope at the sound of a bell ignored the bell after treatment with chlorpromazine. Limited clinical trials were also conducted. Six months later Laborit was also sent samples, which he immediately tested on patients undergoing surgery. In the report that he and his colleagues published in February 1952, they noted that, when the drug was used in situations not related to surgery, it caused a strangely "disinterested" state in patients.

Recognizing that something of critical interest to psychiatry was in the offing, Jean Delay and Pierre Deniker, psychiatrists at the Sainte-Anne mental hospital in Paris, asked Rhône Poulenc to send samples. Delay, the son of a surgeon practicing in Bayonne, France, was a broadly educated individual, having earned degrees in medicine, literature, and philosophy from the Sorbonne in Paris. At the time of these clinical trials he also held a clinical chair in mental disease at the Sorbonne. He had already published influential works in psychiatry as well as several novels. Shortly afterward he wrote a critically acclaimed psychobiography of the contemporary French writer André Gide, who had become his friend; this biography was later translated into English and published in the United States. In this same period, among his many other honors, he was elected to the French Academy. A Parisian by birth, Deniker

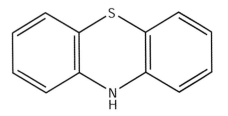

Figure 1. Phenothiazine.

received his M.D. from the Faculty of Medicine (Paris) and was also serving on its faculty when the chlorpromazine trials were in progress.

Patients at Sainte-Anne's were given the drug by itself, not in a mixture, and in successive doses. Delay and Deniker gave a glowing report of the drug's powers, including its effectiveness in subduing the hallucinations and delusions of psychotic patients. "The psychological reaction to the drug is the apparent emotional detachment of the patients, as well as their slower reaction to external stimuli, the reduction of initiative and anxiety without loss of alertness or intellectual functioning." Shortly after, in December 1952, Rhône Poulenc introduced chlorpromazine to the French medical community as "Largactil" (large activity) for a wide variety of purposes, including applications in surgery and psychiatry as well as to counteract pain, morphine addiction, nausea and vomiting, and convulsions.

Meanwhile, Smith, Kline and French, a Philadelphia-based pharmaceutical company (now GlaxoSmithKline), was seeking a cooperative venture with some European pharmaceutical company. Then in early 1952 Rhône Poulenc, which had previously been diffident to overtures from Smith, Kline and French, suddenly displayed great interest in cooperating, because they were just in the process of developing chlorpromazine and hoped to license it in the U.S. market. Smith, Kline and French researchers set to work putting the drug through animal testing and clinical trials that involved some two thousand doctors and their patients in the United States and Canada. Once the U.S. Food and Drug Administration approved chlorpromazine in 1954, for

■ Paul Charpentier.

use in psychiatry and to inhibit nausea and vomiting, Smith, Kline and French began marketing the drug as Thorazine. Within eight months of its appearance on the market it had been administered to over two million patients.

Thanks to these early antipsychotic medications, the number of patients in U.S. mental institutions dropped dramatically: from a high in 1955 of 559,000 inpatients, it was down to 452,000 a decade later. There were, to be sure, drawbacks to the drugs' administration—not least, the potential after long-term use of developing tardive dyskinesia, or distorted involuntary movements (still a problem with today's antipsychotics). On balance, though, the relief of human misery was extraordinary. In terms of scientific progress reserpine and Thorazine also proved to be bonanzas, opening up research into the interaction of brain chemicals and drugs.

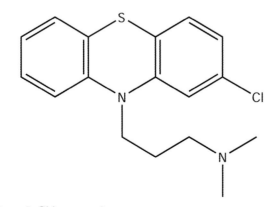

**Figure 3.** Chlorpromazine.

■ Simone Courvoisier.

■ Henri Laborit. Courtesy Albert and
Mary Lasker Foundation.

■ Jean Delay. Photo from P. Pichot, "Jean Delay, 1907–1987," *La Presse Medicale* 29:16 (12 Sept. 1987), 1390–1391. © Masson Editeur.

■ Pierre Deniker. Courtesy Naoline Deniker.

# LEO STERNBACH (1908– ) AND LOWELL RANDALL (1910– )

In the 1950s scientific work on mood-altering medications was booming. Chlorpromazine, an antipsychotic licensed by Smith, Kline and French in 1952 and marketed two years later as Thorazine, revolutionized the treatment of the mentally ill. Its effectiveness and popularity opened the door for further research in this area and led to the development of tranquilizers for a wide variety of clinical applications. Two of the most significant of these new medications were Librium (chlordiazepoxide) and Valium (diazepam). These medications are still prescribed for many people suffering from psychosomatic conditions and anxiety but not from psychoses. Both of these "wonder drugs," developed by Leo Sternbach and Lowell Randall, are products of Hoffmann–La Roche.

Leo Sternbach was born on 7 May 1908 in Abbazia, formerly part of Austria and now part of Croatia. His father was a pharmacist and encouraged his son to follow in his footsteps, although the young Sternbach's interests lay in chemistry. In 1929 Sternbach received his master's degree in pharmacy from the University of Kraków in Poland. He stayed on at the university, receiving his doctorate two years later in organic chemistry.

Sternbach remained at the University of Kraków as a research assistant, but he left in 1937; as a Jew he was unlikely to be promoted at the Polish university. He received a grant to study colloid chemistry and chose to study in Vienna with Wolfgang Pauli (recipient of the 1945 Nobel Prize in physics). Sternbach, who became disinterested in colloids, spent half his days in the organic chemistry laboratory of Sigmund Fränkel. However, his work with Fränkel also failed to hold his interest, so Sternbach moved to the Eidgenössische Technische Hochschule (ETH) in Zurich. There, Sternbach worked with Leopold Ruzicka (recipient of the 1939 Nobel Prize in chemistry).

In 1940 Sternbach accepted an industrial position with Hoffmann–La Roche and moved to Basel. He attempted to synthesize riboflavin (also known as vitamin B$_2$), work that he continued when the anti-foreigner sentiment brought on by World War II forced the company to transfer him to a laboratory in the United States. Shortly after he arrived in New Jersey, Sternbach also began to work on synthesizing biotin, which functions in the metabolic processes that lead to the formation of fats and the utilization of carbon dioxide.

In the 1950s Hoffmann–La Roche wanted to enter the tranquilizer market. Miltown (meprobamate), a Wallace Laboratories tranquilizer that derived from an antiseptic, was already on the market and was immensely popular, and other pharmaceutical companies wanted to capitalize on its success. Sternbach was assigned the task of finding a suitable group of compounds for this work.

Sternbach began looking for a class of compounds that promised to lead to biologically active products and that could provide a number of variations. He recalled working in Kraków with heptoxdiazines, the results of trying to transform amino ketones to make dyes; so he decided to try a variation, benzoheptoxdiazines, because they were relatively unknown but readily available. Sternbach had a hunch that by modifying the structure he could create a biologically active compound. By 1955 he had not yet succeeded, and so he switched gears. He began to work on antibiotics, which were also of interest to Hoffmann–La Roche. Sternbach still worked on tranquilizers in his free time, however, and in 1957 he submitted a sample to Lowell Randall, the chief of pharmacology, for testing.

Lowell Randall was born in North Georgetown, Ohio, on 11 September 1910. He received his B.S. from Mount Union College in 1931 and four years later completed his Ph.D. in biochemistry at the University of Rochester. Soon after, Randall joined the State Civil Service in Massachusetts as an assistant chemist; he then left in 1939 to become a pharmacologist at Burroughs-Wellcome and Company. He remained there until 1946, when he joined Hoffmann–La Roche.

Randall conducted several preliminary tests on the new compound, chlordiazepoxide (a benzodiazepine), using mice and a cat as subjects. These tests focused on muscle relaxation, sedation, and anticonvulsant properties, and the results were compared with similar properties in meprobamate (Miltown), chlorpromazine, and phenobarbital. Chlordiazepoxide was unusual: in the tests it had a tranquilizing effect on the animals, yet they remained alert. It was found to be more potent than the extremely popular Miltown. This compound eventually became Librium, which was introduced in 1960.

Sternbach's success with Librium did not deflect him from continuing his search for suitable tranquilizers. His group began to investigate variations of chlordiazepoxide

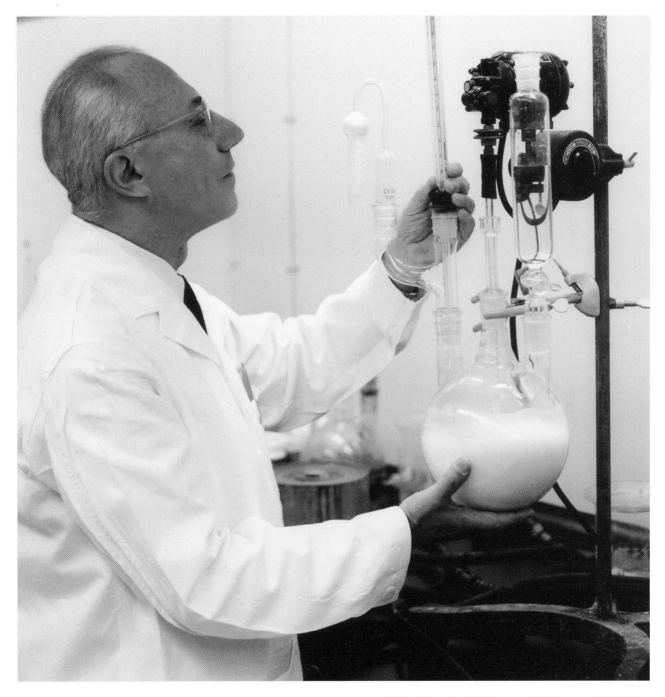

■ Leo Sternbach. Courtesy Hoffmann–La Roche.

and synthetic methods for producing the compounds so discovered. In 1959 the search yielded diazepam, trade named Valium. More effective than Librium at smaller dosages, Valium was introduced in 1963 and continues to be among the most commonly prescribed medications.

Although at the time of its initial development little was known about how Valium works, today we know that it depresses the central nervous system without significantly affecting the autonomic nervous system. Its popularity stems in part from the low occurrence of side effects. When used for extended periods of time, however, it can cause dependence.

After his success with tranquilizers Sternbach rose through the ranks of research at Hoffmann–La Roche, becoming section chief in 1965 and director of medicinal chemistry in 1967. He retired in 1973, becoming a consultant.

Randall retained his position as director of pharmacology at Hoffmann–La Roche until his retirement in 1975.

■ Lowell Randall. Courtesy Hoffmann–La Roche.

# 9. Newer Vaccines

At the end of the eighteenth century Edward Jenner introduced the first vaccine. By using a cowpox virus (at that time considered to be a weaker version of smallpox) that had been passed from cow to man and thereby weakened (or attenuated), Jenner produced a smallpox vaccine. Almost a century later Louis Pasteur (see p. 6) developed a method of attenuating microbes by transferring them from culture to culture at periodic intervals. When these attenuated microbes were injected into the human body, they caused the production of antibodies (which in turn produced immunity) without causing the full-blown disease.

It was this same principle that in the 1950s and 1960s led to the production of vaccines for measles, mumps, and rubella and to one of the vaccines for the dreaded polio. During this period, known as the beginning of the "modern vaccine era," vaccines were developed by very different methods than in Pasteur's time, not only because clinical trials were more elaborate but also because culturing disease-carrying microbes in the laboratory had expanded possibilities. (Biotechnology has, to be sure, taken vaccine development to new levels. See William J. Rutter, p. 193.)

Measles, before the development of the vaccine, was a disease prevalent in children. Though typically not lethal, it was so common—four million cases annually in the United States in the early 1960s—that the number of deaths from such complications as pneumonia far exceeded those from polio. Mumps, too, was a common childhood disease, with an estimated two hundred thousand children in the United States contracting mumps each year during the 1960s. Side effects, however, were potentially more destructive: for example, mumps can cause sterility in postpubescent males. Rubella was less common, with fewer than fifty-eight thousand reported cases in 1969 in the United States. However, when rubella is contracted by pregnant women, it can cause devastating birth defects. But it was polio—the least common of these diseases, with twenty-five to fifty thousand new cases in the United States each year during the early 1950s—that was the most feared, for it could cause permanent paralysis or even death. For most children in the United States today these diseases are not a threat. Widespread vaccination has virtually terminated the spread of the diseases and in the case of polio has even eradicated it from the Western Hemisphere.

# Jonas Salk (1914–1995) and Albert Bruce Sabin (1906–1993)

In the first half of the twentieth century, summer was a dreaded time for children. Although they could enjoy the long summer days of unfettered play, it was also known as "polio season." Children were among the most susceptible to paralytic poliomyelitis (also known as infantile paralysis), a disease that affects the central nervous system and can result in paralysis. When exposed to a poliovirus in the first months of life, infants usually manifested only mild symptoms because they were protected from paralysis by maternal antibodies still present in their bodies. However, as hygienic conditions improved and fewer newborns were exposed to the virus (which is present in human sewage), paralytic poliomyelitis began to appear in older children and adults who did not have an infant's benefit of immunity. President Franklin Delano Roosevelt is perhaps the most famous victim of the poliovirus. In 1921, at the age of thirty-nine, he contracted the disease, one of the thousands that were afflicted that year.

In the early 1950s, twenty-five to fifty thousand new cases of polio occurred each year. Jonas Salk became a national hero, when he allayed the fear of the dreaded disease with his polio vaccine, approved in 1955. Although it was the first polio vaccine, it was not to be the last; Albert Sabin's oral vaccine, introduced in the United States in the 1960s, replaced it. Although the disease was finally brought under control because of these vaccines, the science behind them fired debate that continues to this day.

Jonas Salk was born in New York City on 28 October 1914, his parents' eldest son. His mother was a Russian Jewish immigrant and his father the son of Jewish immigrants. Salk was encouraged throughout his youth to succeed academically. He graduated from high school at the age of fifteen and then entered the City College of New York. Although he originally intended to pursue law, Salk became interested in medicine and altered his career path, graduating with a degree in science in 1933.

At nineteen Salk enrolled in the New York University School of Medicine. His intention was not to practice medicine, however; he wanted to be a medical researcher. Toward the end of his medical education Salk began to work with Thomas Francis, Jr., who was to be his mentor for many years. Salk received his M.D. in 1939 and, after completing his internship at Mt. Sinai Hospital, accepted a National Research Council fellowship to work at the University of Michigan. There he rejoined Francis (who had since moved to Michigan) and spent six years researching the influenza virus and developing a flu vaccine, work largely supported by the U.S. Army. The vaccine that they ultimately developed in 1943 was a killed-virus vaccine: it contained a formalin-killed strain of the influenza virus that could not cause the disease but did induce antibodies able to ward off future viral attacks. Francis and Salk were among the pioneers of the killed-virus vaccine. Up to then, weakened live viruses were used to produce vaccines.

In 1947 Salk accepted a position at the University of Pittsburgh School of Medicine to establish a Virus Research Laboratory. There he devoted his efforts to creating a first-class research environment and to publishing scientific papers on a variety of topics, including poliovirus. Salk's work drew the attention of Basil O'Connor, president of the National Foundation for Infantile Paralysis (now known as the March of Dimes). Salk was invited to participate in a research program sponsored by the foundation. He agreed and took up his assignment of typing polioviruses, a task he credited with building his foundation of knowledge on the virus.

In 1951 the National Foundation typing program confirmed that there were three types of poliovirus. By that time Salk was convinced that the same principle of the killed-virus vaccine he had used to develop an influenza vaccine would work for polio. He also believed that it would be less dangerous than a live vaccine: if the vaccine contained only dead virus, then it could not accidentally cause polio in those inoculated. One difficulty, however, was that large quantities of poliovirus were needed to produce a killed-virus vaccine because a killed virus will not grow in the body after it is administered the way a live virus will. In 1949 John Enders, Thomas Weller, and Frederick Robbins had discovered that poliovirus could be grown in laboratory tissue cultures of non-nerve tissue (earning them the Nobel Prize in physiology or medicine in 1954). The work of Enders and his colleagues paved the way for Salk, for it provided a method of growing the virus without injecting live monkeys.

Salk developed methods for growing large quantities of

■ Jonas Salk surrounded by test tubes. Courtesy family of Jonas Salk.

the three types of polioviruses on cultures of monkey kidney cells. He then killed the viruses with formaldehyde. When injected into monkeys, the vaccine protected them against paralytic poliomyelitis. In 1952 Salk began testing the vaccine in humans, starting with children who had already been infected with the virus. He measured their antibody levels before vaccination and then was excited to see that the levels had been raised significantly by the vaccine.

In 1954 a massive controlled field trial was launched, sponsored by the National Foundation for Infantile Paralysis. Almost two million children in the United States, between the ages of six and nine, participated. In some areas of the country half of these "Polio Pioneers" received the vaccine, while half received a placebo. In other areas of the country children who did not receive any vaccine were carefully observed. On 12 April 1955 Dr. Francis, Salk's mentor and the director of the trial, reported that the vaccine was safe, potent, and 90-percent effective in protecting against paralytic poliomyelitis.

In order to conduct these massive trials, Salk's vaccine needed to be produced on a large scale. Accomplishing this required the assistance of the pharmaceutical industry, and well-known companies like Eli Lilly and Company, Wyeth Laboratories, and Parke, Davis and Company agreed to make the new vaccine.

In the meantime a live-virus vaccine for polio was being developed by Albert Sabin. Sabin, like many scientists of the time, believed that only a living virus would be able to guarantee immunity for an extended period.

Albert Sabin was born on 26 August 1906 in Bialystok, Russia (now Poland). At the age of fifteen he emigrated with his family to the United States. After Sabin graduated from high school in Paterson, New Jersey, his uncle agreed to finance his college education, provided that Sabin study dentistry. After two years preparing for dentistry at New York University, Sabin switched to medicine, having developed an interest in virology. In doing so, he lost his financial support, but odd jobs and scholarships enabled him to continue his education. Sabin received his B.S. in 1928 and afterward enrolled in the New York University College of Medicine.

While at medical school Sabin spent time researching

pneumonia, developing an accurate and efficient method of determining its cause in individual cases—either pneumococcus or virus. He received his M.D. in 1931 and, after completing his internship, traveled to the Lister Institute of Preventative Medicine in London to conduct research. A year later he returned to the United States, having accepted a fellowship at the Rockefeller Institute for Medical Research. There Sabin developed an interest in poliovirus. In 1936 he and a colleague were able to grow poliovirus in brain tissue from a human embryo.

During World War II, Sabin left his polio research to serve in the U.S. Army Medical Corps. There he investigated other diseases like insect-borne encephalitis and dengue, working on vaccines for both.

After the war Sabin accepted a position at the University of Cincinnati College of Medicine as a professor of research pediatrics. He then was able to return to his polio studies. To learn as much as possible about the disease, he and his colleagues performed autopsies on everyone within four hundred miles of Cincinnati who had died of polio. These autopsies indicated that poliovirus affected both the intestinal tract and the central nervous system. From this finding Sabin was able to prove that polio first attacked the intestinal tract before moving on to nerve tissue. This discovery suggested that the virus could be grown in non-nerve tissue, a feat later accomplished in tissue culture by the Nobel laureates Enders, Weller, and Robbins. Growing poliovirus in non-nerve tissue culture was more practical than Sabin's previous achievement of growing it in brain tissue from embryos.

Around the same time that Salk began his work on a killed-virus vaccine, Sabin began work on an attenuated live-virus vaccine. Sabin felt that an oral vaccine would be superior to an injection, as it would be easier to administer. He began to grow and test many virus strains in animals and tissue cultures and eventually found three mutant strains of the virus that appeared to stimulate antibody production without causing paralysis. Sabin then tested these strains on humans: his subjects included himself and his family, research associates, and prisoners from the nearby Chillicothe Penitentiary.

Because Salk's vaccine was being used successfully in the United States, Sabin was not able to get support for a

■ Albert Sabin demonstrates how the oral vaccine for polio is given to children. Courtesy Hauck Center for the Sabin Archives, Cincinnati Medical Heritage Center.

large-scale controlled field trial like the Francis trial of Salk's vaccine. In 1957 Sabin convinced the Soviet Union's Health Ministry to conduct field studies with his vaccine. After the Soviet trial succeeded in 1960, the U.S. Public Health Service approved the vaccine in 1961 for manufacture in the United States, and the World Health Organization began to use live-virus vaccine produced in the U.S.S.R.

In the late 1950s Sabin entered into an agreement with Pfizer to produce his live-virus vaccine. He presented the company with the master strains of the virus, and Pfizer

■ Salk's discovery of the polio vaccine made headlines around the world. Courtesy March of Dimes Birth Defects Foundation.

began to perfect its production technique in its British facilities.

Sabin's live-virus, oral polio vaccine (administered in drops or on a sugar cube) soon replaced Salk's killed-virus, injectable vaccine in many parts of the world. In 1994 the World Health Organization declared that naturally occurring poliovirus had been eradicated from the Western Hemisphere owing to repeated mass immunization campaigns with the Sabin vaccine in Central and South America. The only occurrences of paralytic poliomyelitis in the West after this time were the few cases caused by the live-virus vaccine itself.

During his lifetime Sabin staunchly defended his live-virus vaccine, refusing to believe any evidence that it could cause paralytic poliomyelitis. Salk, for his part, believed that killed-virus vaccine produced equivalent protection in individuals and in communities without any risk of causing paralysis. Despite Sabin's belief the risk of live-virus vaccine-associated paralysis does exist, although it is slight. In 1999 a federal advisory panel recommended that the United States return to Salk's injectable, killed-virus vaccine because it cannot accidentally cause polio.

Although he was the first to produce a polio vaccine, Salk did not win the Nobel Prize or become a member of the National Academy of Sciences. An object of public adulation because of his pioneering work, he spent his life trying to avoid the limelight but nevertheless endured the animosity of many of his colleagues who saw him as a "publicity hound." In 1962 he founded the Salk Institute for Biological Studies in La Jolla, California, an enterprise initially funded with support from the March of Dimes. Salk's own research continued, most significantly on multiple sclerosis, cancer, and AIDS. Salk spent the later years of his life committed to developing a killed-virus vaccine to prevent the development of AIDS in those infected with human immunodeficiency virus.

Sabin, too, continued his work and held a series of influential positions at such organizations as the Weizmann Institute of Science, the U.S. National Cancer Institute, and the National Institutes of Health.

# Maurice Hilleman (1919– )

During his career Maurice Hilleman developed nearly three dozen vaccines for viruses ranging from measles and mumps to hepatitis B and Marek's disease (a visceral leukosis that affects chickens). His research has resulted in the virtual eradication of some of the most feared childhood diseases. Hilleman, often called the "godfather of the modern vaccine era," has been one of the most prolific virologists of the twentieth century. Yet his name is virtually unknown to the public.

Maurice Hilleman was born in Miles City, Montana, on 30 August 1919 and raised on the family farm. He attended Custer County High School but had little inclination to attend college: after graduating, he took a position at a nearby J. C. Penney's store. He did not stay there long, however; when his older brother returned home from seminary for a visit, he insisted that the young Maurice go to college.

Hilleman enrolled at Montana State College, where in 1941 he completed his degree in bacteriology and chemistry. He then proceeded to graduate school at the University of Chicago. In 1944 Hilleman finished a prizewinning dissertation on chlamydia, a parasite that causes such illnesses as conjunctivitis (pinkeye), respiratory tract infections, and sexually transmitted diseases. He received his Ph.D. in bacteriology and parasitology, having completed a five-year program in only three years.

Although his professors urged him to enter academe, Hilleman accepted a position at E. R. Squibb and Sons in New Brunswick, New Jersey. There he continued his research on chlamydia while also beginning his vaccine work, developing vaccines for influenza and Japanese B encephalitis.

Although he did well at Squibb and began to move up the management ladder, Hilleman decided to return to research. In 1948 he accepted a job with the Walter Reed Army Institute of Research. There he studied the adenovirus, which causes such maladies as sore throats and fever, and the influenza virus.

While Hilleman was at Walter Reed, he was courted by Merck. Max Tishler (see p. 97) particularly wanted to hire him and guaranteed him research freedom. Nonetheless Hilleman was unsure that he wanted to go back to industry. Each time he hedged, however, the monetary offer increased; when the salary was high enough to ensure that he would be able to send his daughter to private school, Hilleman accepted.

He moved to the Merck Institute for Therapeutic Research in West Point, Pennsylvania, in 1957, becoming the director of virus and cell biology research. There Hilleman began work on a measles vaccine. He had little confidence in the efficacy of a killed-virus vaccine and so tried a live-virus vaccine. His vaccine was tested at the Children's Hospital of Philadelphia and found effective, but only if administered to children over nine months of age; in younger children maternal antibodies prevented the vaccine from working. Unfortunately, the vaccine also had several adverse side effects, including fever and rash; these reactions were so severe as to be unacceptable. Joseph Stokes of the Children's Hospital had a practical solution to this problem, however. He found that if he injected one arm with human immune gamma globulin—a blood protein that produces antibodies—and then immediately injected the measles vaccine into the other arm, he weakened the vaccine-induced measles enough to prevent the adverse reactions.

A larger trial was conducted in Haverford Township, Pennsylvania, where Stokes's solution worked well. Shortly thereafter, however, it was discovered that the chick embryo cultures used to produce the vaccine were infected with chicken leukemia, which is a potentially carcinogenic virus in humans. Hilleman began his search for hens immune to chicken leukemia, from which he could produce "clean" eggs. He found them at Kimber Farms, which bred experimental hens. Though at first the farm did not want to give up their prized birds, they were eventually convinced of the importance of Hilleman's quest and sold them.

In 1963 the U.S. government licensed the vaccine, trade named Rubeovax. The gamma globulin was named Gammagee. Merck was the sole producer of the attenuated virus vaccine (see Louis Pasteur, p. 6, on attenuation). Merck continued to improve the vaccine, however, eventually reducing its side effects and eliminating the need for the gamma globulin. This improved version was marketed as Attenuvax.

In the meantime Hilleman turned his attention to mumps, a virus that causes inflammation of the salivary glands, thereby making chewing and swallowing difficult. In the 1950s a killed-virus vaccine for the disease had been

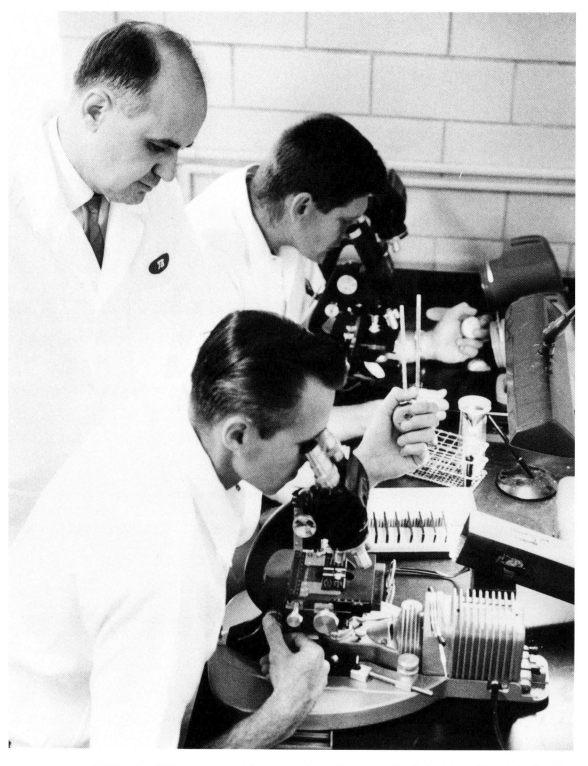

■ Maurice Hilleman supervising researchers. Courtesy Merck Archives, Merck & Co., Inc.

developed, but it only provided temporary protection. In 1963 Hilleman's six-year-old daughter, Jeryl Lynn, developed the then-common childhood malady. Hilleman took blood and saliva samples from her and transported them to the lab, where he succeeded in isolating the strain (thereafter known as the Jeryl Lynn strain of mumps) and attenuating it in chick embryos.

Two years later Hilleman conducted a trial at Children's Hospital. The researchers tested several mumps vaccines, which had been created using varying levels of virus attenuation. They found a successful vaccine, trade named Mumpsvax, which was licensed in 1967 by the U.S. government.

Meanwhile work began on the rubella virus, also known as German measles. Although rubella is a rather mild virus, if a pregnant woman contracts the disease in the early months of pregnancy, it can cause birth defects. In 1948 Sir Frank Macfarlane Burnet (who would receive the 1960 Nobel Prize in physiology or medicine) succeeded in providing "passive immunity" to pregnant women who had been exposed to rubella. Burnet used gamma globulin from individuals who had had the disease to create the temporary immunity, which succeeded in preventing birth defects. The focus then began to shift to providing long-term protection against the disease. In 1960 Thomas Weller and an Army research group from Walter Reed succeeded in isolating the rubella virus. Two years later Hilleman and his colleagues at Merck isolated a strain of the virus, known as the Benoit strain. They decided to focus on developing a killed-virus vaccine—which they thought could be produced and tested quickly—because of a predicted outbreak of rubella in the United States (it occurred between 1963 and 1965). In 1965, however, Hilleman and his team shifted to a live-virus vaccine.

Their first challenge was to find a suitable substance in which to grow the virus. Chick embryo cells, which they had used to grow both the mumps and the measles viruses, were not susceptible to rubella. They decided to try duck embryo cells, and they succeeded on their first attempts. Hilleman and his team were ready to produce a vaccine.

In the meantime the Division of Biological Standards of the National Institutes of Health (NIH) was also working on a rubella vaccine. They used kidney cells from cattle and a different strain of rubella (known as HPV-77) and suc-

ceeded in producing a live-virus vaccine in 1966. Given this success, Mary Lasker, an influential lobbyist for health issues, suggested to Hilleman that Merck discard their Benoit strain in favor of the NIH's HPV-77. Rather than abandon his work, Hilleman decided to continue research on *both* strains. He soon discovered that the HPV-77 virus was too toxic and further attenuated it in the duck embryo cells, creating "HPV-77 duck." Tests showed that this new attenuated version was as effective as the Benoit strain, and so Merck continued with HPV-77 duck alone. Initial clinical trials of the rubella vaccine were promising, and in 1969 the NIH's Division of Biological Standards licensed the Merck rubella vaccine, known as Meruvax.

These three major vaccines—for measles, mumps, and rubella—did their job well. Instances of these diseases decreased significantly. However, there was still concern that as the diseases came under control, parents would neglect to immunize their children. In order to decrease this likelihood, Merck tried to simplify the vaccination process, creating combination vaccines. Specifically, Hilleman and his team sought to create a combination vaccine that would combat measles, mumps, and rubella. This task was particularly difficult because they needed to ensure that the side effects were minimal, that the potency was guaranteed, and that the three vaccines did not interfere with each others' efficacy.

Merck succeeded in its effort to create a combination vaccine. Clinical trials were conducted by the Department of Pediatrics at the University of Pennsylvania Medical School. The trials were successful, and the government licensed the so-called MMR vaccine in 1971. While this research was going on, Merck also developed several two-virus vaccines, including one for measles and mumps, one for mumps and rubella, and one for measles and rubella.

Today, thanks to Hilleman's vaccines, measles, mumps, and rubella are under control. His career at Merck was long and distinguished, and his research did not end with MMR. In 1978, for example, he introduced Pneumovax, a vaccine for pneumococcal pneumonia. Four years later came Heptavax-B, a vaccine for hepatitis B that was derived from human plasma. In the mid-1980s, after developing nearly three dozen vaccines, Hilleman retired from Merck.

# 10. Clinical Testing
## and
## Reform of the
## 1906 Pure Food and Drugs Act

Today we think of clinical trials in terms of federal regulation,

but such trials did not become a firm part of the federal drug

approval process until 1962, almost twenty years after Austin

Bradford Hill ushered in the era of the controlled clinical trial.

His emphasis on scientific methodology, combined with the

work of Walter Campbell and Frances Kelsey in the U.S. Food

and Drug Administration, led to the development of modern

food and drug legislation.

The 1906 Pure Food and Drugs Act (see Harvey Washington Wylie, p. 26) pioneered pure food and drug legislation. The 1906 act established outer boundaries, making adulteration and misbranding illegal. But as time wore on, it became evident that this landmark bill was far from comprehensive. It did not, for example, address standards for safety or efficacy—the effectiveness of a drug for its advertised purpose. It did not penalize false advertising, and cosmetics did not fall under its purview.

The 1938 Food, Drug, and Cosmetic Act provided for some of the omissions of the 1906 act. It addressed issues relating to advertising and product quality. It required manufacturers to prove the safety of drugs prior to marketing. And it brought not only cosmetics but also medical devices under the control of the FDA. However, no provisions were made for efficacy.

Efficacy remained an issue until the 1962 Kefauver-Harris Amendments, named for Senator Estes Kefauver of Tennessee, who introduced the bill, and Representative Oren Harris of Arkansas, the Chairman of the House Committee on Interstate and Foreign Commerce. This bill forced companies to demonstrate the effectiveness of new drugs as well as to send adverse reaction reports to the FDA. In addition, it specified that well-controlled clinical trials were necessary for new drugs requesting FDA approval. It also required that the permission of patients be obtained before enrolling them in clinical studies.

Today drug companies must go through a complex procedure to gain FDA approval for their products (see chart on facing page), which in recent years has been much expedited. The FDA approves over fifty new drugs each year. After a drug has been developed—a process that typically includes multiple phases of animal testing and clinical testing on humans and that takes years to conduct—a company files a new drug application (NDA) with the FDA. The NDA can have as many as fifteen individual sections and must include information about the drug's safety and effectiveness as well as data from the animal studies and clinical trials. Once the application is filed, it is reviewed by some or all of the six departments of the FDA's Center for Drug

**The new drug development process.** The process can be accelerated in a carefully regulated way for new therapies intended to treat people with life-threatening and severely debilitating illnesses, especially where no satisfactory alternatives exist.

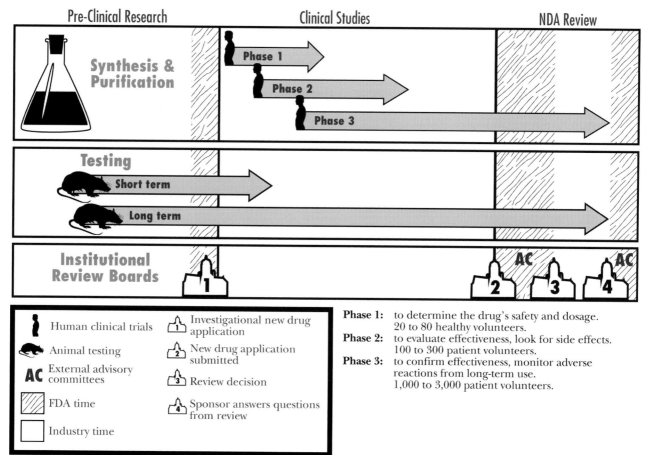

Pre-Clinical Research    Clinical Studies    NDA Review

Synthesis & Purification

Phase 1

Phase 2

Phase 3

Testing

Short term

Long term

Institutional Review Boards

**Legend:**
- Human clinical trials
- Animal testing
- **AC** External advisory committees
- FDA time
- Industry time
- ① Investigational new drug application
- ② New drug application submitted
- ③ Review decision
- ④ Sponsor answers questions from review

**Phase 1:** to determine the drug's safety and dosage. 20 to 80 healthy volunteers.
**Phase 2:** to evaluate effectiveness, look for side effects. 100 to 300 patient volunteers.
**Phase 3:** to confirm effectiveness, monitor adverse reactions from long-term use. 1,000 to 3,000 patient volunteers.

Evaluation and Research (depending on the application): medical, statistical, biopharmaceutical, pharmacological, chemical, and microbiological. In addition, outside advice is sought from an advisory committee. Further information may be requested of the company, and revisions to the application may be required. The instructions for the use of the drug are reviewed, and manufacturing sites and clinical testing sites are inspected. The FDA then decides whether to approve the drug. In the past it sometimes took up to eight years after filing the NDA to gain FDA approval, but today this period has been reduced to an average of one year.

## WALTER CAMPBELL (1877–1963)

For decades after it was passed, the 1906 Pure Food and Drugs Act (see Harvey Washington Wiley, p. 26) was held up as a model piece of legislation. However, as time wore on, it became increasingly evident that there were difficulties with enforcement and that the United States needed a stronger, more encompassing law. Spurred on by Walter G. Campbell, chief of the U.S. Food and Drug Administration (FDA), the Federal Food, Drug, and Cosmetic Act was passed in 1938.

Walter Campbell was born in Kentucky and attended the University of Kentucky, earning a B.A. in 1902. Three years later he received a law degree from the University of Louisville. In 1907 Campbell took the Civil Service examination in order to become an inspector for the enforcement of the 1906 Pure Food and Drugs Act, then administered by the U.S. Department of Agriculture (USDA). Shortly thereafter Harvey Wiley handpicked Campbell for the position of chief inspector.

When Wiley retired from the USDA in 1912, Campbell was offered the position of chief of the Bureau of Chemistry. He refused, however, strongly believing that such a position would be best filled by a chemist, not by a lawyer. In 1914, when the district system was instituted, Campbell became chief of the Eastern District, and by 1917, he was assistant chief of the USDA. In 1927 Campbell moved to the position of chief of the newly created Food, Drug, and Insecticide Administration (later known as the Food and Drug Administration).

Throughout his career at the Department of Agriculture and the FDA, Campbell was extremely dedicated to enforcing food and drug legislation. Through his work he began to notice the shortcomings of the landmark 1906 law and the need for new legislation. For example, the 1906 law did not address the growing cosmetics industry. Nor did it adequately control false advertising of drugs, food adulteration, or the so-called "patent medicines"—popular tonics, typically with a high alcohol content, that claimed to cure a variety of ailments and were widely available, with or without a prescription.

Related legislation had been passed since 1906, including the Sherley Amendment, enacted in 1911, which prohibited medicine labels from carrying false claims; the Gould Amendment of 1913, which required that the quantity of contents be included on food labels; and the 1930 McNary-Mapes Amendment, which established quality standards for canned food. This legislation, though, was not sufficient to address all of the inadequacies of the 1906 act.

Public acknowledgment of the need for new legislation was beginning to increase. In response to consumer discontent, Consumer's Research, an organization that became highly active in consumer-related legislation, was founded in 1929. In addition, muckraking books, including Arthur Kallet and F. J. Schlink's *100,000,000 Guinea Pigs* (1932) and M. C. Phillips's *Skin Deep: The Truth about Beauty Aids* (1934), proliferated during the 1930s, detailing precisely how consumers were being cheated and describing the dangerous nature of some drugs, foods, and cosmetics.

In 1933 Campbell and Rexford Guy Tugwell, the newly appointed assistant secretary of agriculture and former member of Franklin Delano Roosevelt's Brain Trust, agreed to revise the 1906 bill. Tugwell quickly attained President Roosevelt's approval for the undertaking, and the work began. The following June a draft bill was introduced to Congress. Dubbed the "Tugwell Bill" by its opponents (Tugwell being generally unpopular and highly criticized for his socialist leanings), it was sponsored by Senator Royal Copeland of New York.

Meanwhile Campbell and his office began to publicize the need for new legislation. They distributed information as widely as possible, arranging for speaking engagements and radio spots and mass mailing pertinent articles. Perhaps their most successful tool was an exhibit known as the "Chamber of Horrors." This series of posters graphically depicted the victims of the shortcomings in the 1906 bill and became extremely popular. Reproductions were made and loaned to interested organizations. Manufacturing and other industrial pharmaceutical organizations opposing the new legislation forced Campbell and his colleagues to cease their education efforts, citing the 1919 Deficiency Appropriations Act, which prohibited federal agencies from using funds for lobbying. Education efforts did eventually resume but on a much smaller scale.

Many thought that the proposed bill gave too much discretionary power to the FDA. After several revisions it was

■ Walter Campbell. Courtesy FDA History Office, Rockville, Maryland.

reintroduced in February 1934. Although women's groups and professional medical and pharmaceutical organizations still supported the legislation, it was not popular enough to pass. Opposition from industrial organizations remained too strong.

A number of other legislative attempts would be made before a new bill would pass. To compound the difficulties, Tugwell resigned his position in 1936; Campbell lost not only a strong supporter but also his main communication route to President Roosevelt.

In the face of yet another failure, a drug-related tragedy spurred politicians and industry leaders to work together on a new law. In 1937 the Samuel E. Massengill Company of Bristol, Tennessee, introduced an elixir of sulfanilamide (see Gerhard Domagk, p. 65), a liquid version of the wonder drug used to treat a variety of infections. Unfortunately no toxicity tests had been made on this new medication, which used the poisonous diethylene glycol as a solvent, and 107 people died after taking the drug. When news of the first few deaths reached Campbell, he amassed his entire field force to deal with the problem. Recalls and public announcements were made, and Campbell's staff tried to track down every last drop of the fatal drug. Under the 1906 act the

only violations with which Campbell could charge Samuel E. Massengill related to misbranding. The situation led industrial interests to realize the necessity of new legislation, and public outrage demanded that a bill be passed quickly.

In December 1937 a new FDA-drafted bill with stringent food and drug manufacturing and testing regulations was introduced in the Senate. Almost simultaneously a similar measure was introduced into the House. This latter measure was preferred for its clarity and was supported by the FDA. Its provisions were merged into the previous Copeland bill, drafted in 1936, and the legislation was passed in the spring of 1938.

Overall, Campbell was pleased with the 1938 Food, Drug, and Cosmetic Act. It vastly improved the 1906 act in terms of both regulation and enforcement. It brought both cosmetics and medical devices under the control of the FDA and increased penalties for adulteration and fraud. In addition, the FDA was given the authority to establish standards for quality. However, despite its efforts, the FDA did not gain control over advertising; those powers were given to the Federal Trade Commission.

Campbell remained at the FDA until 1944, establishing the enforcement plans for the 1938 act.

# SIR AUSTIN BRADFORD HILL (1897–1991)

Before a drug can be approved by the FDA, it must undergo a rigorous set of clinical tests that prove the drug's safety and effectiveness. The development of the controlled clinical trial, which seeks to eliminate bias in drug testing through the use of randomization, is largely credited to Sir Austin Bradford Hill, a British statistician.

Austin Bradford Hill was born on 8 July 1897, the third of six children. His father, Leonard Hill, was a well-known research physiologist, first at the London Hospital and later at the Medical Research Committee (subsequently the Medical Research Council [MRC]).

Hill grew up in Loughton, Essex, and attended the Chigwell School. He had always intended to follow in his father's footsteps and become a doctor. However, by the time he completed his secondary education, World War I had broken out, and Hill volunteered to serve in the Royal Naval Air Service. There he learned to fly planes, and he was eventually posted to the Greek island of Imbros, as part of a force providing support for the Allied attack on the Dardanelles.

A misdiagnosed and incorrectly treated attack of tuberculosis, not the dangerous nature of his service, sent Hill home to England. Since there seemed little hope that he would recover, he was awarded a 100-percent disability pension for life. He surprised everyone by getting well, though only after a long convalescence. During this time it became evident that Hill would be unable to perform the strenuous activities required by the sciences. So he chose to study the most scientific of the social sciences—economics. In 1922 he received his B.Sc. from the University of London, after taking all of his courses by correspondence.

On the recommendation of Major Greenwood, a long-time family friend and the Senior Statistical Officer to the Ministry of Health, Hill was given a grant by the MRC to investigate the relationship between migration within the United Kingdom and mortality rates. He soon earned a permanent position on the statistical staff of the MRC and expanded his studies into the field of occupational health. He remained at the MRC until 1927, when Major Greenwood was appointed professor of epidemiology and vital statistics at the new London School of Hygiene and Tropi-

cal Medicine (LSHTM). Major Greenwood took Hill along with him, and LSHTM was to be Hill's professional home for the remainder of his career.

At LSHTM, Hill's responsibilities broadened to include teaching. Fortunately, he was a natural teacher. Throughout his career he was known and respected for his clarity of expression and interest in simplicity. One of the so-called Hill's laws stated that a project that could not be explained to a child was probably not worth doing. This emphasis on simplicity was characteristic not only of his lectures and his clinical studies but also of his writing. In 1937 Hill was persuaded by the *Lancet* to publish a series of his lectures, a series that evolved later that year into his landmark book *Principles of Medical Statistics*.

Hill's statistical experiences led him to believe in the necessity of randomized clinical trials. Clinical trials are used to determine the effectiveness of a new treatment or to compare a new treatment with a commonly used treatment. In each case, patients are separated into two groups. One group receives the new treatment while the other group receives either a placebo or the more common treatment, depending on the purpose of the trial. In some cases a clinical trial is double blinded—that is, neither the doctor nor the patient knows which of the two treatments the patient is receiving. Randomization is a statistical method that takes the element of chance into account and thereby provides more accurate results. Because patients have physiological differences, the patient groups used in a clinical trial must be selected randomly. This method helps to ensure that each patient group includes a variety of individuals, thereby representing the whole population more accurately. Hill believed that randomization was essential for the comparison of different medical treatments, for it would eliminate bias—conditions that would unconsciously favor one outcome over another—in a study.

Hill was not the first to apply the statistical method of randomization to scientific experiments. Sir Ronald Aylmer Fisher (1890–1962), a British geneticist, is also considered a pioneer in this area. Fisher used randomization in plant-breeding experiments in order to make the selection of experimental materials more objective.

■ Sir Austin Bradford Hill. Courtesy London School of Hygiene and Tropical Medicine.

In 1946 Hill succeeded in convincing the MRC to use randomization in two separate trials. One of these trials was of a whooping cough vaccine. The other trial was for the use of streptomycin—an antibiotic recently discovered by Selman Waksman (p. 90)—in treating pulmonary tuberculosis. The streptomycin trial was organized by the MRC Tuberculosis Research Unit and coordinated by Marc Daniels, to whom Hill credited the trial's success. This trial marks the beginning of the use of modern controlled clinical trials.

During this time Hill also became involved in a study of the connection between cigarette smoking and carcinoma of the lung. An increasing number of deaths were being caused by lung cancer. In 1947 the MRC convened a conference to discuss possible reasons for this increase. Many theories were considered, including that of cigarette smoking. Working with Sir Richard Doll of the MRC, Hill designed the questionnaire for this long-term pioneering study and provided the statistical expertise, while Doll provided the clinical expertise. Their case-control study was carefully designed. Initial interviews were conducted with cancer patients in twenty London hospitals. The interviewers, who did not know what type of cancer each patient had, asked about smoking habits. Patients and controls were matched for age, gender, and length of hospital stay. The results of this study showed that smoking and lung cancer were closely related. Although these results were controversial at the time, subsequent studies have confirmed their findings. After this seminal clinical trial Hill and Doll continued to collaborate on clinical trials, primarily on extended studies of the connection between smoking and lung cancer.

By the early 1950s the randomized controlled trial was beginning to be accepted in certain sectors of the research community. In 1962 the U.S. government specified that prior to federal drug approval "adequate and well-controlled investigations, including clinical investigation," were required. The Kefauver-Harris Amendment, as this legislation was known, expanded the pure food and drug legislation already in place (see Harvey Washington Wiley, p. 26) and provided greater support for Hill's methodology.

Austin Bradford Hill's work has had another far-reaching effect. His research helped to establish that chronic diseases were as suitable subjects for statistical study and treatment as infectious diseases.

Hill became a Fellow of the Royal Society in 1954 and was knighted seven years later.

# FRANCES OLDHAM KELSEY (1914– )

It is estimated that ten thousand children worldwide have been born deformed because of the use of the sedative thalidomide during pregnancy. This tragic situation inspired the 1962 Kefauver-Harris Amendment, strengthening drug regulation by forcing companies to disclose side effects and prove efficacy. Today thalidomide is being investigated for use in a variety of conditions, including inflammation, severe canker sores in AIDS patients, several forms of cancer, and macular degeneration, a condition that can result in blindness. For the most part, though, thalidomide has remained off the market in the United States for almost forty years, owing primarily to the tenacity of one FDA medical officer, Frances Kelsey.

Frances Oldham Kelsey was born on 24 July 1914 in Cobble Hill, British Columbia. She was educated in Canada, receiving her B.Sc. from McGill University in 1934 and her M.Sc. from the same institution a year later. She moved to the United States to continue her education at the University of Chicago, earning her Ph.D. in pharmacology in 1938 and her M.D. in 1950. In 1937, while in graduate school, Kelsey worked on some of the experiments that helped to identify diethylene glycol as the toxic agent in the deadly elixir of sulfanilamide (see Walter Campbell, p. 136). She taught at both the University of Chicago and the University of South Dakota before becoming a general practitioner in Vermillion, South Dakota.

Kelsey joined the U.S. Food and Drug Administration (FDA) in August 1960. A month later she became a reviewing medical officer in the Bureau of Medicine, with the task of reviewing new drug applications. The first application she received was from the Cincinnati-based William S. Merrell Company (now Aventis Pharma) for thalidomide.

Developed in the 1950s by the German company Chemie Grünenthal, thalidomide was considered to be an unusually safe sedative with few side effects and low toxicity. It was soon found to reduce the nausea and vomiting associated with pregnancy. Beginning in 1956, it was marketed under the trade name Contergan and became widely used in Europe. Grünenthal, wishing to capitalize on the drug's success, began to license the manufacture and distribution of thalidomide to companies in other countries. The American licensee was the Merrell Company.

The Merrell Company submitted its application for FDA approval of thalidomide (under the brand name Kevadon) in September 1960. Under law at that time the FDA had sixty days to decide whether a drug was safe—no provisions for determining or testing efficacy were made. The applications were reviewed by three individuals: a chemist, a pharmacologist, and a medical officer. If the FDA did not respond within sixty days, the drug manufacturer could assume it was approved and could then market the medication.

When Kelsey and the other two FDA officials reviewed the thalidomide application, they quickly developed questions about the drug. For example, the pharmacologist on the team, Oyam Jiro, felt that the company's toxicity tests had not been sufficient. The chemist, Lee Geismar, questioned the company's manufacturing controls. Kelsey herself felt that the safety data were inconclusive and questioned why large amounts of the drug—ostensibly a sedative—could be given to animals without producing drowsiness.

As a result Kelsey sent the Merrell Company a letter within sixty days stating that their application was incomplete. In mid-January the company resubmitted their application, but questions remained about the drug's metabolism, excretion and absorption levels, and toxicity. At the end of February, Kelsey read a letter by A. Leslie Florence in the *British Medical Journal* that noted a connection between thalidomide use and the development of peripheral neuritis—a condition characterized by inflammation of nerves, particularly in the arms and legs, causing pain, muscle weakness, and a decrease in reflexes. Although the European producers of thalidomide had discovered this connection earlier, the American company was not required to submit such information in their FDA application.

This *British Medical Journal* article also led Kelsey to suspect that thalidomide could damage nerves in a fetus. She had an interest in teratogens—drugs that can create malformations in fetuses—and the FDA and the American Academy of Pediatrics had been concerned about the effects of drugs taken during pregnancy. In November 1961 the first case of phocomelia was reported and linked to thalidomide use. Phocomelia is a severe birth defect that causes the malformation of the fetus's limbs. Subsequent findings showed that phocomelia was most likely to occur

■ Frances Kelsey receiving the President's Distinguished Federal Civilian Service Award from President John F. Kennedy, in 1962. Courtesy FDA History Office, Rockville, Maryland.

in cases in which the mother took thalidomide during the first trimester of her pregnancy, when the limb buds of the fetus are forming.

Clinical testing of the drug ceased (except for clinical studies on its possible use in cancer), and the Merrell Company withdrew their thalidomide application from the FDA. Given the potential danger of the medication, the FDA tracked down the remaining samples in the United States. During this search the FDA found a few cases of phocomelia that had resulted from clinical trials in the United States.

These discoveries had little press coverage until the *Washington Post* reporter Morton Mintz broke the story in July 1962. This publicity led to increased public support for stronger drug regulation. The result was the Kefauver-Harris Amendment, passed in October 1962. This amendment required that companies demonstrate the effectiveness of new drugs, that companies send adverse reaction reports to the FDA, and that patients participating in clinical studies give consent. The provision for efficacy was retroactively applied to 1938, when the Food, Drug, and Cosmetic Act was passed. In order to test the effectiveness of drugs that had been approved on the basis of safety between 1938 and 1962, the FDA contracted with the National Academy of Sciences and the National Research Council.

In 1962 Kelsey received the President's Distinguished Federal Civilian Service Award from President John F. Kennedy. She remained at the FDA and in 1995 at the age of eighty-one received yet another appointment there as deputy for scientific and medical affairs in the Office of Compliance of the FDA's Center for Drug Evaluation and Research.

# PART III.

# THE 1960S TO THE 1980S

**Clockwise from top left:** Alfred Alberts (courtesy Merck Archives, Merck & Co., Inc.); Arthur Patchett (courtesy Merck Archives, Merck & Co., Inc.); Miguel Ondetti (courtesy Bristol-Myers Squibb Company); Fu-Kuen Lin (courtesy Amgen, Inc.); Gertrude Elion and George Hitchings (courtesy GlaxoWellcome Inc. Heritage Center).

# 11. Recent Cardiovascular and Antiulcer Drugs

Up until the 1960s most drugs used to treat cardiovascular problems and ease the pain of ulcers were discovered without benefit of knowing exactly how or why they worked at the cellular or molecular level. Such was the case for digitoxin, the derivative of the foxglove plant (Figure) used since medieval times both as a poison and as a corrective for irregular heartbeat. Similarly, how nitroglycerin—introduced into medical practice in 1877 as an emergency treatment for severe anginal or heart pain—works went unexplained until recently. Even today, when a cellular-molecular explanation commonly precedes the development of a drug, such explanations are occasionally revised in the light of new evidence. Still, the approach of targeting cells and molecules within particular physiological processes, or so-called rational design, has proved exceptionally fruitful.

The number of effective cardiovascular and antiulcer drugs available to patients greatly increased in the era of rational design, which began in the 1960s. The widespread prescription of blood pressure–lowering drugs that arose from new, more systematized methods of research has improved the health and longevity of millions. Similarly, cholesterol-reducing medicines are maintaining arteries and hearts in relatively good working order throughout the human lifespan. And thanks to new medicines, stomach and intestinal ulcers, once painful and life threatening, have become well-controlled conditions.

In rational design, molecular targets—substrates or receptors for enzymes, hormones, or neurotransmitter substances known to be involved in a particular disease—are first selected on the basis of knowledge of the body's biochemical and physiological processes. Next, scientists investigate compounds they believe might block the function of the chosen target molecule, and then they settle on a "lead" candidate—often another drug that exhibits to some small degree the activity required. Alternatively, or simultaneously, a search is done among natural sources, including substances produced by various microbes, that will have the desired biochemical effect. Then drawing on the store of molecular structure-activity relationships that chemists have built up over time, scientists modify the lead-candidate molecule to give it greater strength and make it easier to administer as a drug.

Characterizing this period of pharmaceuticals research as the era of "rational design" is controversial. It implies that predecessors acted irrationally, when in fact the notion that enzymes and other critical biological chemicals act at specific cellular and molecular sites has a long history going back to Emil Fischer, Paul Ehrlich (p. 17), Henry Hallett Dale (p. 59), and Daniel Bovet (p. 68), among others. In addition, contemporary research often benefits from sheer chance or moments of intuition. Still, as acknowledged by the 1988 Nobel Prize in physiology or medicine, awarded to James Whyte Black (p. 148) and George Hitchings and Gertrude Elion (p. 170) "for their discoveries of important principles for drug treatment," research in pharmaceuticals has achieved a much more secure scientific footing than ever before. Moreover, the drugs so discovered have undergone more extensive testing for safety and efficacy under the regulation of the 1962 Kefauver-Harris Amendments to the Food, Drug, and Cosmetic Act (see Frances Oldham Kelsey, p. 142).

**Figure.** Foxglove, or digitalis, illustrated here in a sixteenth-century English herbal, was known as the source of a heart medicine for centuries before there was an explanation in modern scientific terms for its efficacy. Rembert Dodoens, *A Nievve Herbal,* trans. Henry Lyte (London, 1578). Courtesy Library of the College of Physicians of Philadelphia.

# JAMES WHYTE BLACK (1924– )

In his pursuit of likely targets for therapeutic intervention James Whyte Black was among the first to use receptor theory rigorously—that is, that there are special receptors for each of the body's chemical messengers that trigger important physiological changes throughout the body. Using this theory, Black discovered a new class of cardiovascular drugs, the beta-blockers, and a class of antiulcer medicines, the histamine-2 (H$_2$)-blockers.

Black was the fourth of five sons of a mining engineer and manager of a colliery in Fife, Scotland. He described himself as coasting through most of his school years, daydreaming except during two periods of intense study—of music between the ages of twelve and fourteen and of mathematics between the ages of fourteen and sixteen. He was, however, able to win a competitive scholarship to St. Andrews University. Although he did not realize it at the time, his family could not have otherwise supported another university student.

Under the influence of an older brother who had also attended St. Andrews, he chose to study medicine and, after graduating, joined the physiology department there as an assistant lecturer. In 1947 he took a teaching position in a medical school in Singapore to earn money to repay the debts incurred by his own medical education. Returning to England in 1950, he was asked to start a new physiology department at the University of Glasgow Veterinary School, where he built up a research laboratory equipped with the most advanced cardiovascular technology he could obtain.

In Glasgow he and his colleagues made preliminary investigations into the two areas in which he would make the most significant contributions: histamine-stimulated gastric acid secretion and the means of increasing the supply of oxygen to the hearts of patients with narrowed coronary arteries. It occurred to Black, though, that there was another approach to remedying the oxygen insufficiency that is experienced by some physically or emotionally stressed cardiac patients and that causes anginal pain or worse, heart attack. That approach was to reduce the heart's demand for oxygen.

Black thought that intervening in hormonal activity might provide the control he was seeking. In the 1940s further research had been conducted on the functions of epinephrine (adrenaline) (see John Jacob Abel and Jokichi Takamine, p. 47) and norepinephrine (noradrenaline), which had been discovered in 1946 by the Swedish physiologist Ulf Von Euler. These hormones each exhibited puzzling, nearly contradictory, effects on various organs in the body, including the heart. In 1948 Raymond Ahlquist of the Medical College of Georgia suggested that such organs possessed two sets of receptors, which he designated alpha and beta, that mediated the action of adrenaline, thus explaining how the same hormone could have quite distinct effects. The alpha-receptors of the heart, for example, use adrenaline to contract the muscle; the beta-receptors relax the muscle but increase the heart rate.

In 1956 Black set out to look for a specific adrenaline antagonist, that is, a substance that would inhibit the heart rate–increasing action of adrenaline that produced a high demand for oxygen. He first sought help from Imperial Chemical Industries (ICI) Pharmaceuticals Division (now Zeneca), and he later joined their new Alderley Park laboratories near Manchester.

Meanwhile, at Eli Lilly in Indianapolis, I. H. Slater and C. E. Powell, who were looking for a new bronchodilator to use in asthmatic patients, synthesized dichloroisoproterenol (DCI). In the course of their testing they noted that adrenaline, long known as a bronchodilator, had no effect on tracheal strips that had previously been exposed to DCI; that is to say, DCI antagonized adrenaline. Two scientists at Emory University in Atlanta—N. C. Moran and M. E. Perkins—then tested DCI's effect on the heart. They found that the drug also inhibited the changes in heart rate and muscle tension produced by adrenaline, seeming to block the beta-receptors of Ahlquist's theory. DCI, however, displayed considerable adrenaline-like activity—a fact that slowed Lilly's own development of DCI as a heart medicine.

After some initial hesitation, in early 1960 Black and his colleagues at ICI took DCI as their lead compound (a promising molecule on which to perform structural modifications). The organic chemist John Stephenson replaced DCI's chlorine atoms with a second benzene ring and thus synthesized pronethalol. Pronethalol proved effective at relieving anginal pain and at regulating some irregularities in patients' heartbeats, but soon biological and clinical studies in

■ James Whyte Black. Courtesy Albert and Mary Lasker Foundation.

animals and man revealed fundamental problems. The drug appeared to penetrate the central nervous system and cause such undesirable effects as dizziness and vomiting. (Later long-term toxicological studies seemed to show that pronethalol also caused cancer in rats—results that no one outside ICI could repeat.) The chemists went to work on developing a new molecule, propranolol. This new drug displayed no such adverse effects, and ICI brought it to market in 1964 as the first widely used beta-blocker. One of the scientists who worked on the clinical development of beta-blockers, Brian Pritchard, a clinical pharmacologist at University College, London, discovered that drugs of this class had additional effects: they reduced high blood pressure, a use for which they are commonly prescribed today, and they could be used for heart conditions.

In 1963, while propranolol was still being launched, Black left ICI for Smith, Kline and French Laboratories (now GlaxoSmithKline) in Welwyn Garden City near London. He was eager to start a new research program, believing that the antihistamines of the day that were used to treat allergies (see Daniel Bovet, p. 68) were analogous to the alpha-receptor antagonists in the cardiovascular system and that the equivalent of a beta-receptor antagonist was needed to block, for example, histamine-stimulated acid secretion in the stomach.

Black and his associates—including Graham Durant, Robin Ganellin, and John Emmett, all chemists, and Michael Parsons, a pharmacologist—set about creating chemical analogues of histamine and testing them in vitro and in vivo. In their first breakthrough they made a molecule that stimulated acid secretion without causing the other histamine responses, thus implying that there is indeed a second receptor, which they labeled $H_2$. Finding an antagonist proved difficult. By 1968, with more than two hundred compounds synthesized, they had not yet discovered an $H_2$-blocker. Next, Parsons developed a more sensitive assay, and they noticed that a compound that Durant had earlier synthesized with a long side chain based on guanidine, a nitrogen-containing compound, now displayed some slight antagonist activity. They then took this compound as their lead and extended its side chain, overcoming many chemical difficulties in the process, and created burimamide (1972).

Since burimamide was not active when taken orally, the team continued their work, changing the electronic properties of the compound by introducing new groups and a sulfur atom in the midst of the side chain. The result, metiamide, was active orally and was ten times more potent than burimamide. Because burimamide and metiamide both contained the potentially toxic thiourea group in the side chain, the chemists, now led by Ganellin, continued their syntheses. They then produced cimetidine, which contains a cyanoguanidine group instead. Meanwhile, of seven hundred patients treated, a few taking metiamide developed agranulocytosis, a reversible blood disorder; so all further development focused on cimetidine. In 1976 in the United Kingdom and in 1977 in the United States, it was launched as Tagamet (derived from *anTAGonist* and *ciMETidine*) and was enthusiastically received by doctors and patients alike. It and similar drugs developed by other companies, like Glaxo's Zantac, transformed the lives of ulcer patients around the world.

Soon the problem became how to produce economically the vast quantities of Tagamet that the market demanded and also how to provide patent protection in those countries that patent only processes but not products. George Wellman and Lee Webb, working out of Smith, Kline and French's Philadelphia base, eliminated a bottleneck step in the original process, the reduction of an imidazole ester intermediate with lithium aluminum hydride. This step was difficult and expensive to perform. Several alternative methods of performing the reduction were devised and patented, but the most cost-effective used sodium in liquid ammonia.

Meanwhile, during the final phases of the development of cimetidine in 1973, Black took up the chair of pharmacology at University College, London. Four years later he moved to the Wellcome Research Laboratories as director of their therapeutic research division, and since 1984 he has been a professor (now emeritus) of analytical pharmacology at Kings College, London. In 1981, seven years before he received his Nobel Prize in physiology or medicine, he was knighted by Queen Elizabeth II.

## Miguel Ondetti (1930– )

In the 1970s Miguel A. Ondetti and his colleagues at E. R. Squibb (now Bristol-Myers Squibb) used the drug discovery principles pioneered by James Whyte Black (p. 148) and George Hitchings and Gertrude Elion (p. 170) to develop a new means of treating high blood pressure, or hypertension. The new drugs were called angiotensin-converting enzyme (ACE) inhibitors. Ondetti started out on the path of discovery a decade after these pioneers and worked on inhibiting an enzyme whose structure had already been investigated. His predecessors were not so well informed about the structures of the molecules with which their research was engaged because X-ray crystallography and other analytical techniques had not advanced that far.

Ondetti began his career in Buenos Aires, Argentina, where he grew up in a family of Italian descent. He and his brother set up a makeshift chemistry laboratory in their home. While attempting to electrolyze copper sulfate, Ondetti suffered a strong electric shock, which put an end to his home experiments. It did not, however, put an end to his desire to become a chemist. But he and his brother attended a commercial high school so that they could quickly obtain relatively well-paying jobs to help support their family.

Once the would-be chemist was secure in his position as a bookkeeper, he applied to the university. He was turned down because he had not taken the required preparatory courses. It was, however, possible to take examinations in the various subjects instead. Ondetti, after studying on his own, took some thirty examinations and was eventually admitted to the University of Buenos Aires. He used to arrive at work in the payroll department of Argentina's Department of Energy at 7:00 A.M. and leave at 2:00 P.M. in order to attend lectures and laboratories.

By 1955 he had earned a licentiate in chemistry (equivalent to a master's degree), after which he won a scholarship to work on his doctoral dissertation at the Squibb Institute for Medical Research in Buenos Aires. In 1950s Argentina, while Juan Pérón was president, Squibb had been given monopoly rights to manufacture antibiotics. In exchange for this privilege, the firm was required to invest some of its profits in the country—hence the institute and Ondetti's scholarship.

After the scholarship year was over, Ondetti was offered a full-time position at Squibb, which he turned down. He wanted to try something else. So he got a job in development at another chemical company. He only lasted a week, though, and returned to his bookkeeping job. Luckily, Squibb soon had another position open, which Ondetti took in 1957, the same year that he received his Ph.D. from the University of Buenos Aires.

At Squibb, Ondetti worked in the natural-products area, isolating biologically active alkaloids from plant materials (see Joseph-Bienaimé Caventou and Pierre-Joseph Pelletier, p. 3). To make ends meet, Ondetti taught chemistry in the evenings at his alma mater and at the Catholic Institute for Teachers.

In 1960 Ondetti was offered a position with Squibb in New Jersey, where the laboratory facilities were superior and the salaries much higher. He accepted. There he was assigned to a newly formed research group led by Miklos Bodanszky, a Hungarian refugee. The group was working on peptides, molecules made up of amino acids in which the amino group of one is united with the carboxyl group of another. Peptide research was a new field for Ondetti, but he learned quickly. In fact, Ondetti was chosen to replace Bodanszky when in 1966 the latter left Squibb to take an academic position.

Among the peptides that Ondetti investigated were insulin; bradykinin, a substance with several functions in the body, including the dilation of certain blood vessels; secretin, a hormone (the first identified as a "hormone"; see John Jacob Abel and Jokichi Takamine, p. 47) that stimulates the pancreas to secrete bicarbonate and water, in other words, that acts as a natural antacid; and cholecystolinin, which stimulates the contraction of the gallbladder and the secretion of enzymes from the pancreas. The purpose of the program was to make new drugs out of some of the body's own peptides. But because peptides can generally be easily broken down by digestive enzymes and therefore must be injected, they proved persistently difficult to develop into the orally administered form that patients find more acceptable.

When Arnold D. Welch became president of the Squibb Institute in 1967, researchers were asked to focus on developing cardiovascular drugs. By this time a great deal was known about the many factors that contribute to high blood pressure, including the role of renin. This substance is produced in the kidneys and cleaves angiotensinogen, a large

■ Miguel Ondetti working with molecular models of captopril. Courtesy Bristol-Myers Squibb Company.

protein manufactured in the liver, to produce angiotensin I. In turn, angiotensin I is cleaved by ACE, which is synthesized mostly in the lungs, to produce angiotensin II. Angiotensin II had been implicated as a cause of constriction of blood vessels, which causes blood pressure to rise. If the converting enzyme could be inhibited, some types of hypertension could then be controlled.

In early 1968 Ondetti was invited to attend a meeting with a consultant, John Vane, then with Britain's Royal College of Surgeons, who was soon to become famous for his explanation of the efficacy of aspirin (it interferes with the production of certain prostaglandins, unsaturated fatty acids that cause painful inflammation). Vane reported that recent results from his laboratory seemed to show that peptides from the venom of the Brazilian pit viper (*Bothrops jararaca*) could block the conversion of angiotensin I to II and meanwhile activate bradykinin, which allowed the bradykinin to do its work of dilating blood vessels. A bite from the viper caused its victim's death through a sharp drop in blood pressure, but these same peptides might prove useful in treating hypertension.

In a project to isolate, characterize, and eventually synthesize these blocking peptides, Ondetti teamed up with David Cushman, a biochemist. Cushman was to develop tests to show the effect of the isolated substances on ACE. Working with Nina Williams, one of his assistants, Ondetti first isolated and determined the structure of a peptide with eleven fragments. Then his longtime assistant Emily Sabo synthesized it under his direction. Bernard Rubin led the pharmacologists involved in the project, whose job was to measure the effects of various likely candidates on isolated smooth-muscle tissue and then on rats.

The most likely candidate was teprotide, a peptide of nine fragments, but however hard they tried, Ondetti's team could not modify the molecule so that it could be taken orally. They then began to screen compounds randomly, looking for evidence of activity against ACE. Two thousand non-peptide compounds later, they were no closer to their goal. Because the research seemed to be at an impasse, in 1973 Squibb's work on ACE inhibitors officially ended. The peptide-synthesis group was subsumed by a new antibiotics group, which Ondetti headed as a step in his assuming ever-broadening responsibilities in the administration of research at Squibb.

But Ondetti and Cushman remained interested in the subject of ACE inhibitors. In 1974 Cushman noticed a paper by L. D. Byers and R. V. Wolfenden, who had found an inhibitor of another enzyme and explained the inhibitor's success as a function of its structure. These researchers deduced that their enzyme inhibitor combined in one molecule certain features of the two products resulting from the action of the enzyme on its normal substrate. Specifically, their inhibitor contained molecular groups similar to those on the fragments of the normal substrate that showed how that substrate had been bound by the enzyme before being split. Armed with this rationale, Ondetti and Sabo synthesized an amino acid molecule, succinyl-L-proline, which they thought would effectively tie up ACE. Right off the bat, they had found a weak but specific ACE inhibitor and bradykinin potentiator.

While still officially assigned to other research projects, the investigators from the old team then proceeded to synthesize and test new molecules. But soon they were able to convince their managers that they were engaged in an important project that deserved their undivided attention and the support of other researchers at the Squibb Institute. Having modified the design concepts that they had originally borrowed from Byers and Wolfenden, they ultimately produced s-3-mercaptopropanoyl-L-proline, or captopril. It showed ten times more activity in vitro than teprotide and in vivo was almost as efficacious when given orally as when given intravenously.

Clinical trials of captopril began in 1976. The rapid escalation of the oral dose, the use of the drug in all sorts of complicated cases, and the consequent side effects led the U.S. Food and Drug Administration to restrict its approval in 1982 for use only as a drug of last resort, after alternatives had been exhausted. In 1984, after further clinical trials, approval was extended for use in all degrees of hypertension.

Following in the wake of captopril—a wake into which many pharmaceutical researchers jumped—Ondetti chose to investigate peptidases, the enzymes that control peptide synthesis, as a better route to finding orally active and more specifically targeted drugs.

# John Duncia (1954– ) and David Carini (1956– )

The search for ways to combat high blood pressure continued even after the invention of angiotensin-converting enzyme (ACE) inhibitors (see Miguel Ondetti, p. 151). Since high blood pressure has few obvious symptoms until a heart attack or stroke occurs, patients tend to stop taking their medicine at the first sign of side effects and then suffer the long-term consequences of their condition. Unwanted side effects of ACE inhibitors, such as a dry persistent cough in some patients, would be eliminated if a drug could be devised that insofar as possible acted only to prevent angiotensin II from triggering the tensing of the smooth muscles lining the blood vessels. Researchers thought they could accomplish this by blocking the specific receptors for angiotensin II on muscle cells. Almost nothing was known about the molecular shape of this receptor, except of course that angiotensin II fit into it. At DuPont's pharmaceutical division (now part of Bristol-Myers Squibb), two young chemists, John Duncia and David Carini, made the key discoveries in the development of losartan (trade named Cozaar), the first of the angiotensin-II antagonists, which are as effective as ACE inhibitors but have fewer side effects.

John Duncia grew up in Detroit. One Christmas his father gave him a colorful chemistry book for children, which he enjoyed immensely. When he took high school chemistry, however, he encountered some difficulties. He could not, for example, understand the concept of the mole until his father, who was a civil engineer, hauled out his own old textbooks and helped his son with this and other problems. Eventually Duncia caught on and took advanced-placement chemistry at Catholic Central High School. Later he majored in chemistry at the University of Michigan at Dearborn and completed a Ph.D. in chemistry at Princeton University.

Albert Carini's father was a chemist who supported his son's interest in science. Another important source of encouragement was Carini's high school chemistry teacher, who designed an independent study course for him to make up for the lack of an advanced-placement chemistry course at his small Vermont high school. Carini received his B.S. degree in 1978 from Rensselaer Polytechnic Institute and his Ph.D. in chemistry in 1982 from the Massachusetts Institute of Technology. He then went to work for the new pharmaceutical group being formed at DuPont.

Shortly after Duncia joined DuPont in 1981, he was placed on the project to discover an angiotensin-II antagonist. He assumed that a fragment of angiotensin II might compete for the same receptor site. He attempted without success to follow Ondetti's example (see p. 151) of designing a molecule by using fragments of the original peptides involved; then from those peptide fragments he planned to create a nonpeptide that could be administered orally. Meanwhile a large-scale project to screen molecules from DuPont's molecular "library" for their effect on blood pressure was under way, but it was not producing results either. The real breakthrough came from another source altogether. Scientists at Takeda Chemical Industries in Japan were looking in their molecular libraries for an angiotensin-II antagonist and discovered and patented a series of rather small nonpeptide molecules. These compounds were originally prepared as diuretics, substances that promote the body's elimination of excess fluids, but instead they displayed some slight potential for reducing blood pressure: the smooth muscles were not contracting as before. Carini was then called on to synthesize a few of these compounds.

Meanwhile, Andrew Chiu, an experienced pharmacologist on the team, set up a means of assaying in vitro whether these compounds were in fact binding to the angiotensin-II receptors. He used radioactively labeled angiotensin II, which then became bound to receptors in isolated smooth-muscle tissue from rats. Next he added the compounds being tested and measured how much of the radiolabeled angiotensin II was displaced. The more potent a compound, the more radiolabeled angiotensin II was displaced. Pancras Wong, another pharmacologist, then tested the compounds' activity compared with that of other substances implicated in producing high blood pressure, for example, epinephrine (adrenaline). Wong found that the Takeda molecules did not block the increase in blood pressure produced by epinephrine or the other hypertensive agents. They only blocked the increase in blood pressure caused by angiotensin II. Thus the Takeda molecules were lowering blood pressure, when given either intravenously or orally at high doses, via the

■ John Duncia (left) and David Carini. Courtesy DuPont Pharmaceutical Company.

selective blockade of the angiotensin-II receptor. It was like being presented with the lead compound on a "golden platter," as Duncia recalled, and the question became how to improve these promising Takeda molecules.

By using some rather simple computer modeling, Duncia superimposed the Takeda molecule that most effectively blocked the angiotensin-II receptors over a three-dimensional molecular drawing of angiotensin II; the goal was to see what chemical groups might be added to the molecule to enlarge it and to make it more closely resemble angiotensin II in the way it binds to the receptor. He especially noted a second carboxylic acid group (COOH) that was present on angiotensin II but not on the Takeda molecule. After he used chemical means to add such a group to the Takeda molecule, Chiu's testing revealed that the new molecule had tenfold greater binding capacity to the angiotensin-II receptor.

The project was gaining momentum as more chemical manipulations and more testing took place to yield compounds that were a hundredfold and then a thousandfold more potent binders than the Takeda molecule. Soon Carini was drawn back on the project. He was to solve the problem of how to make the DuPont molecules orally active, that is, able to get out of the digestive tract and into the bloodstream. The compound contained an amide bond (HNC=O) linking two benzene rings together. The amide group tended to attract water molecules, making it impossible for the drug to slip through the walls of specialized intestinal cells known as greasy cells and enter the bloodstream. Carini eliminated the amide bond and replaced it with a carbon-carbon single bond, which resulted in an orally active system of two phenyl groups (benzene rings). Finally, to achieve the best results in lowering blood pressure, Duncia found that the carboxylic group that he had originally attached to the best Takeda molecule had to be replaced by another acidic group, a tetrazole, which is a five-membered ring of atoms consisting of one carbon and four nitrogen atoms. All told, the work in the research laboratories alone, which was driven by a combination of theoretical principles, experience, intuition, and serendipity, took nearly two years.

DuPont took some steps to develop losartan (Cozaar) for the market, but to finish the job of rapidly putting the compound through clinical studies, the company turned to the more mature development and marketing departments of the Merck organization. The two companies launched in 1991 a joint venture—the DuPont Merck Pharmaceutical Company. Then in 1998 DuPont bought out Merck's shares to create DuPont Pharmaceuticals, which Bristol-Myers Squibb subsequently purchased.

In the spring of 1995 the U.S. Food and Drug Administration approved losartan for the treatment of hypertension, and it is now marketed by Merck as Cozaar. Subsequently, other pharmaceutical companies developed angiotensin-II antagonists, among them Bristol-Myers Squibb and Smith-Kline Beecham (now GlaxoSmithKline).

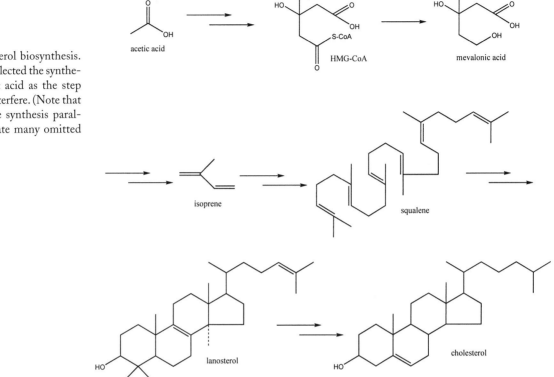

**Figure.** Cholesterol biosynthesis. Alfred Alberts selected the synthesis of mevalonic acid as the step with which to interfere. (Note that elsewhere in the synthesis parallel arrows indicate many omitted steps.)

## ALFRED W. ALBERTS (1931– ), GEORG ALBERS-SCHÖNBERG (1929– ), AND ARTHUR A. PATCHETT (1929– )

Cholesterol is a critical component of cell membranes, and steroid hormones and bile acids are synthesized from it in our bodies. Yet we are reminded daily by advertisements and items in the news about low-density lipoprotein (LDL), or "bad," cholesterol and its harmful deposition in our arteries, which can lead to heart attacks and strokes. Only about 30 percent of the cholesterol in the human body comes from the food we eat; the remaining 70 percent is synthesized by the body itself. This synthesis takes place mainly in the liver, starting with acetic acid, which can be derived from a wide variety of sources. Clearly, modifying dietary intake of cholesterol can only partially alleviate the cholesterol problem that many people face—much of it owing to hereditary flaws in their biochemistries. In the 1970s researchers at Merck and Company discovered that lovastatin (trade named Mevacor), the product of a mold, could intervene effectively in the *bio*synthesis of cholesterol.

When they began their research, the Merck scientists already knew a great deal about the steps of this synthesis because of previous research in the field. The synthesis begins with the condensation of acetic acid into isoprene units via the key intermediate of mevalonic acid; then the pathway, leaving out details, leads from squalene to lanosterol to cholesterol (Figure). It was Merck's own Karl Folkers who discovered mevalonic acid in 1956. At every step of the way researchers found enzymes and also coenzymes, organic substances that are required, in addition to an enzyme and a substrate, for an enzymatic reaction to proceed.

In 1973 Michael Brown and Joseph Goldstein, researchers at the University of Texas Health Sciences Center in Dallas, solved an additional piece in the puzzle of cholesterol metabolism. They discovered LDL receptors on the surface of cells; these serve as conduits for cholesterol, which is taken out of the bloodstream and then used in the cell or

■ Alfred Alberts in the laboratory in 1963. Courtesy Merck Archives, Merck & Co., Inc.

■ Arthur Patchett at the blackboard in 1962. Courtesy Merck Archives, Merck & Co., Inc.

■ Georg Albers-Schönberg, circa 1990. Courtesy Merck Archives, Merck & Co., Inc.

excreted. They were also able to explain why an early anticholesterol drug, cholestyramine (which Merck scientists had discovered in the 1950s), actually worked—for a while. It stimulated the increase of LDL receptors in the liver, thus increasing the amount of LDL excreted as bile acid, but unfortunately it also triggered an increase in the liver's synthesis of cholesterol. An obvious target for researchers became the cholesterol synthesis itself.

In pursuing this line of research in the 1970s, Merck sought the advice of P. Roy Vagelos, an expert on the metabolism of fats (lipids) and at the time a faculty member at Washington University in St. Louis. In 1975 Vagelos joined Merck as senior vice president in pharmaceutical research, a position from which he rose to become Merck's chairman

and chief executive officer. Vagelos brought with him his longtime associate Alfred Alberts. Alberts, while still in graduate school, was originally hired to assist Vagelos at the National Institutes of Health (NIH). A native of Brooklyn, Alberts had completed a bachelor's degree in biology at Brooklyn College in 1953, had served in the army, and was then completing course work for a Ph.D. in zoology at the University of Maryland. One of his professors at Maryland, Earl Stadtman, a noted enzymologist and Vagelos's chief at NIH, announced in class that there were some openings for laboratory workers at NIH, and Alberts leaped at this opportunity. The work with Vagelos turned out to be so interesting that Alberts never returned to Maryland to complete his dissertation.

When Alberts arrived at Merck, he set about putting together an in vitro assay that focused on a key step early in the cholesterol synthesis. This reaction is in turn controlled by a feedback loop that is in part dependent on the level of LDL cholesterol in the bloodstream. Alberts planned to screen microbial extracts for substances that would inhibit the action of an enzyme critical to the synthesis of mevalonic acid.

Meanwhile Akira Endo at Sankyo Laboratories in Japan was working on a similar project. In 1976, after screening thousands of microbial extracts, Endo announced that he had discovered a substance—mevastatin, isolated from the mold *Penicillium citrinum*—that reduced cholesterol production by acting on the same step in the biosynthesis as Alberts's target. Mevastatin was also isolated by scientists at Beecham Research Laboratories in England as a potential antifungal, compactin.

Information arriving from both the Japanese and the English researchers confirmed Alberts's belief that he had taken the right track, which inspired the Merck team to redouble their efforts. Alberts then turned to Arthur Patchett, director of Merck's New Lead Discovery department, to provide microbial broths maintained in Merck's fermentation products for screening (FERPS) system that could be submitted to the assay he had designed to find an appropriate enzyme inhibitor. Patchett had been at Merck for twenty years, having joined the company soon after receiving a Ph.D. from Harvard University in organic chemistry. At Harvard he had worked with Robert Burns Woodward (p. 114) on synthesizing lanosterol and understanding its role in the biosynthesis of cholesterol. Almost simultaneously with the work that he was doing with Alberts, Patchett was working on enalapril (Vasotec), Merck's angiotensin-converting enzyme (ACE) inhibitor (see Miguel Ondetti, p. 151).

Working with Patchett was Richard Monaghan, who made several contributions to the development of Merck's new anticholesterol drug—not least his decision to preserve in FERPS a mold broth that turned out to contain a substance that would serve Alberts's purpose. The mold *Aspergillus terreus* had been cultured in Merck's laboratories in Spain with the hope that it would produce a substance that

could control a common protozoan infection in chickens, but extracts from the mold broth did not perform as expected back at headquarters in New Jersey. Normal procedure would have been to discard the culture, but Monaghan thought it interesting and saved it.

In late September 1978, Alberts was ready to begin his assay. His assistant Julie Chen prepared special glass tubes containing mixtures of the enzyme that Alberts wanted to inhibit and its radioactively tagged substrate—reactants that would normally produce mevalonic acid. The substrate was hydroxymethylglutaryl coenzyme A, or HMG-CoA, and the enzyme was, appropriately, HMG-CoA reductase. Chen then placed microbial extracts in the tubes—a single extract in each one. If an extract had no effect on the reaction, radioactive mevalonic acid would appear and could be separated out. If, however, the extract inhibited the reaction, little or no radioactive mevalonic acid would be found. First Alberts's group tested some hundred samples from FERPS that had been used in other assays, and mevalonic acid was produced in the presence of each of them. Next Patchett's group sent over twenty or so new extracts, and surprisingly the tube with the eighteenth sample produced no mevalonic acid whatsoever. After ruling out accidents—such as leaving out the enzyme—Alberts was delighted to conclude that there was indeed a substance in this extract that would interrupt the body's synthesis of cholesterol.

But was it the same substance that the Japanese and English investigators had already discovered? To determine this, Alberts needed the assistance of expert analytical chemists. Carl Hoffman from Patchett's laboratory, who had participated in the mevalonic acid research under Folkers twenty years before, isolated the active principle from the extract. Georg Albers-Schönberg set to work determining the molecular structure of the isolated substance from its mass spectra and nuclear magnetic resonance (NMR).

The child of a German father and a Swiss mother, Albers-Schönberg received his secondary and higher education in Switzerland during and shortly after World War II. After earning a doctorate in organic chemistry from the University of Zurich, he accepted a postdoctoral fellowship at the Massachusetts Institute of Technology. At MIT he carried out the synthesis of natural products, deploying the new physical methods of structural determination, including NMR techniques. At the end of this fellowship, in 1965 Albers-Schönberg accepted a position at Merck Research Laboratories. Among the projects at Merck to which he contributed shortly before his work on lovastatin was the development of ivermectin, an effective treatment for river blindness, a disease that used to infect and blind millions of people living in tropical countries around the world.

Albers-Schönberg was brought the newly isolated active principle that prevented the production of mevalonic acid on a Friday afternoon. He and his associates were scheduled to make a report on the material's structure at a meeting on the following Monday, so they worked furiously through the weekend. Mass spectra showed right away that the structure was very similar to compactin's. Indeed, NMR techniques showed that the Merck molecule differed from compactin by only a single additional methyl group. But as subsequent studies showed, that group had a big effect on the potency of lovastatin, making it two to three times more powerful than compactin in reducing cholesterol levels.

The long weekend in the laboratory paid off. Although Endo also discovered this more powerful molecule, a U.S. patent for lovastatin was granted to the Merck scientists in 1980. The drug was approved by the U.S. Food and Drug Administration in 1987 after rigorous clinical testing. Since that time other anticholesterol drugs have come on the market, and Merck itself has produced a semisynthetic version of lovastatin, trade named Zocor.

# 12. Recent Antihistamines and Antidepressants

Great advances were made in allergy medicines and antidepressants once researchers had learned enough about various specialized receptors that recognize and respond to the body's own chemical messengers. Although allergies and depression might not seem to be related types of medical conditions, the medicines prescribed to alleviate these conditions are in fact closely related.

Daniel Bovet (p. 68) recognized early on that histamines and other chemicals related to the autonomous nervous system affect the brain. Similarly, the first truly antipsychotic drug, chlorpromazine (Figure), was developed by Paul Charpentier and his colleagues (p. 118) when they were trying to produce a better antihistamine.

Scientists worked to improve the early antihistamines and antidepressants because the drugs tended to block more than one receptor in the nervous or other systems, thus causing unpleasant and sometimes dangerous side effects. Meanwhile, James Whyte Black (p. 148) demonstrated that at least some of the body's own chemicals, such as adrenaline and histamine, fit several receptors that respond quite differently. Working in this vein, scientists since the 1960s have been able to design antihistamines that do not cause drowsiness. In a similar fashion, they have produced antidepressants that target serotonin, and they are also at work trying to produce antipsychotics with fewer side effects.

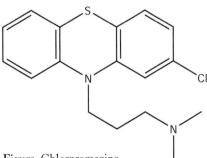

**Figure.** Chlorpromazine.

# Albert A. Carr (1930– )

Allergies affect more than fifty million Americans, and most of us are probably familiar with the symptoms: sneezing, itching, red eyes, and runny nose, to name a few. Today many allergy sufferers depend on such well-known drugs as Claritin, Zyrtec, and Allegra to control allergic outbreaks. These drugs are part of a second generation of nonsedating antihistamines, compounds that block at the cellular level the reception of histamine—the substance that triggers allergy symptoms—without causing drowsiness or sedation. Seldane, the first of the first generation of nonsedating antihistamines, was discovered in 1973 by Albert Carr, who also discovered Allegra.

Albert Anthony Carr was born in Covington, Kentucky, in the midst of the Great Depression. His family moved across the Ohio River to Cincinnati when he was about a year old. In high school, English and the arts had little appeal for him, but he liked the sciences—things that had predictability—and did well in them. Carr decided that he wanted to pursue medicine, and he received a scholarship to study at Xavier University, a local Catholic university. Carr began with a pre-med curriculum but soon added extra chemistry courses to his class load in order to earn a bachelor's degree in chemistry as well as a pre-med degree. Even then, Carr felt that his education in chemistry was incomplete, so he stayed on at Xavier to earn a master's degree in organic chemistry.

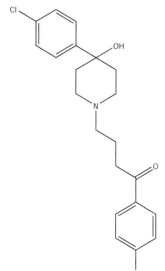

**Figure 1.** Haloperidol.

Albert Carr thus left behind his early intentions of becoming a physician, but he never abandoned his objective of improving health care. He entered a Ph.D. program at the University of Florida at Gainesville and accepted a fellowship to develop anticancer compounds in the Cancer Research Laboratory there. To prepare for this research, Carr took a number of biochemistry and biology courses. This combination left him well positioned for making major contributions to the pharmaceutical industry.

In 1958 Carr was hired by the William S. Merrell Company (now Aventis Pharma) in his hometown of Cincinnati. He plunged right into work on drugs that affect the central nervous system. He focused on antipsychotic drugs, using as a starting point an antipsychotic much more powerful than chlorpromazine, which a Belgian company had just discovered—haloperidol (see Paul Charpentier et al., p. 118). Like most antipsychotics, haloperidol interacted in multiple ways with the nervous system and other systems. It was known to block dopamine receptors in the brain, increasing the supply of dopamine available to the nerves there and presumably producing antipsychotic effects. It also produced such undesirable effects as orthostatic hypotension, which is the sudden draining of blood from the brain when a person stands up after lying down, and tardive dyskinesia, a serious drug-induced motor disorder, usually characterized by repetitive involuntary movements. These problems spurred further research to find better antipsychotics. Meanwhile, more and more was being learned about neuroreceptors and the complex roles in drug function played by the body's natural neurotransmitters, such as dopamine, serotinin, acetylcholine, adrenaline, and noradrenaline. Carr spent years working out the relationships between the molecular structures of substances that could mimic these transmitter substances and thus selectively block their receptors.

Carr also attempted to capitalize on the antihistaminic properties of most antipsychotics—properties that operate mainly outside the central nervous system. Antihistamines are substances that compete with histamines for receptors. They prevent rather than reverse histamine actions. Histamines were discovered in 1911 by Sir Henry Hallett Dale (p. 59) and antihistamines in 1937 by Daniel Bovet (p. 68) and Anne Marie Staub. Histamine, the substance naturally

■ Albert Carr in 1999. Photograph by James Traynham.

released from body tissue in response to injury or irritation, has distinct activities. It expands blood vessels and increases the permeability of vessel walls to facilitate the transfer of components of the immune system. As a result histamine causes swelling that indicates that the body is fighting an infection. It is also the culprit that causes the common discomforts associated with allergies. These attacks occur when the immune system is hypersensitive to foreign antigens (dust, pollen, and mold are typical irritants to overly sensitive immune systems), which elicit the release of histamine and the accompanying symptoms of an allergic reaction.

The first commercial antihistamine, Antergan, was mar-

keted in 1942, and the door to developing more specific, less toxic, and stronger antihistamines was opened. All of the earliest antihistamines made those who took them drowsy, so much so that some of them were and are used as sleeping aids in such preparations as Tylenol PM and Benadryl.

In the midst of the project to develop antihistamines, Carr turned to a "library" that Merrell had collected of previously synthesized substances that are active in the central nervous system. Focusing on those that totally lacked antihistaminic properties, he began to work on modifying the antipsychotic drug haloperidol (Figure 1) and other

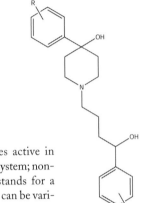

**Figure 2.** Molecules active in the central nervous system; non-antihistaminic. R stands for a group of atoms that can be variously specified.

trographic evidence suggested that in the body two oxygen atoms were added to each molecule of terfenadine. There are over twenty different positions where the additional atoms could be added to the structure. Instead of synthesizing all these structures, Carr surmised that the oxygen atoms were in fact present as the oxidation of just one of the methyl groups on terfenadine (see Figure 3). Named fexofenadine, this molecule was also a nonsedating antihistamine and is now sold under the name Allegra. Tested in conjunction with erythromycin and ketoconazole, Allegra does not cause the harmful side effects of its predecessor, Seldane.

Figures 1 through 3 show the progression of Carr's research from the antipsychotic agent haloperidol and the nonantihistaminics in Merrell's library of molecules that are active in the central nervous system to the nonsedating antihistamines Seldane and Allegra. Although not appreciated at the time, the addition of a third phenyl ring to the structures shown in Figures 1 and 2 (see the ring at the top of Figure 3) is now believed to be responsible both for the antihistaminic properties of Seldane and Allegra and for the molecules' inability to cross the blood-brain barrier, so that they do not cause drowsiness.

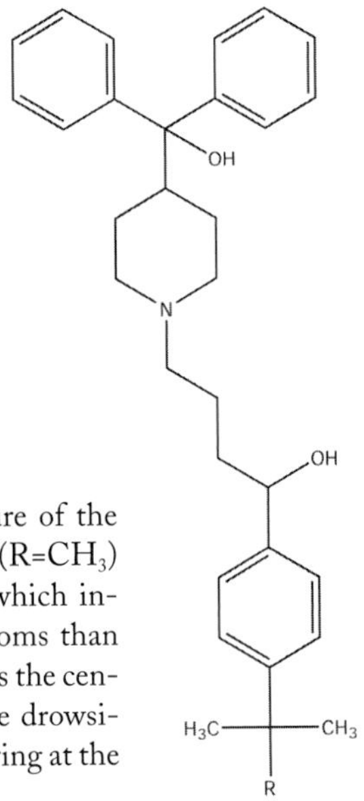

molecules from the library (Figure 2). By the time he finished, he had a new molecule that carried a total of three phenyl (benzene) groups (Figure 3). The new molecule now exhibited antihistaminic activity, and radioactive labeling showed that it did not cross the blood-brain barrier. This barrier is formed by the capillaries in the brain, which are less permeable than other capillaries in the body. Molecules must pass through both the capillary wall and the glial cells that wrap the capillaries in order to reach the nerve cells of the brain; consequently, only water, oxygen, carbon dioxide, and most but not all molecules soluble in fats and oils can pass through. Thus, Carr's discovery was of a *nonsedating* antihistamine, since it did not permeate the brain. He had successfully accomplished what others in the pharmaceutical industry had tried to do for over thirty years. Named terfenadine, the drug was given the trade name of Seldane and became available as a prescription drug in 1985.

After more than a decade terfenadine was removed from the market by the U.S. Food and Drug Administration: it became linked to potentially serious heart rhythm abnormalities when taken in conjunction with a number of other commonly prescribed drugs, such as the antibiotic erythromycin and the antifungal ketoconazole. Earlier, following common precedent in pharmaceuticals development (see Robert L. McNeil, Jr., p. 13, and Daniel Bovet, p. 68), Carr had set out to isolate terfenadine's metabolite, the substance into which the drug was converted in the body. Mass spec-

**Figure 3.** The basic structure of the two antihistamines Seldane ($R=CH_3$) and Allegra ($R=COOH$), which includes two more oxygen atoms than Seldane. Neither drug affects the central nervous system to cause drowsiness. Note the third phenyl ring at the top right of the structure.

■ Left to right: David Wong, Ray Fuller, and Bryan Molloy. Reprinted with permission.

# RAY W. FULLER (1935–1996), DAVID T. WONG (1935– ), AND BRYAN B. MOLLOY (1939– )

Since its introduction by Eli Lilly in 1988, Prozac (fluoxetine) has been hailed as a wonder drug, a safe antidepressant that is more effective than those of previous eras. Though questions have been raised about Prozac's side effects and true level of efficacy, it has undeniably become one of the most-prescribed antidepressants, with worldwide sales remaining in the $2 to $3 billion range even after a decade on the market and increased competition from other pharmaceutical companies. Prozac has enabled many of the estimated ten million people in the United States suffering from depression to lead normal lives. The development of this groundbreaking drug was undertaken by three Eli Lilly researchers: Ray Fuller, David Wong, and Bryan Molloy.

During the 1950s a class of antidepressants known as tricyclics was developed. Although effective, these drugs also cause a variety of undesirable side effects, including dizziness, blurred vision, and constipation. The next generation of antidepressants was known as monoamine oxidase inhibi-

tors, or MAOIs. Unfortunately, MAOIs are highly toxic. Further research was needed to develop an effective antidepressant with low toxicity and a low incidence of side effects.

A giant breakthrough in antidepressant research occurred in 1974, when Eli Lilly scientists reported on their studies of fluoxetine, the first in a class of antidepressants known as selective serotonin reuptake inhibitors, or SSRIs.

Most antidepressants work in the same manner: they increase the amount of neurotransmitters in the brain. Neurotransmitters are the chemicals released by nerve cells to stimulate neurons, thereby passing impulses through the nervous system. It is believed that depression is especially affected by levels of the neurotransmitters serotonin (first isolated from blood in 1948) and norepinephrine (discovered in the mid-1940s by Ulf von Euler, later a Nobel laureate). Antidepressants differ in respect to which neurotransmitters they increase and how they do it. MAOIs inhibit the enzyme monoamine oxidase from destroying the neurotransmitters

167

norepinephrine, serotonin, and dopamine, thereby increasing the levels of all three. Tricyclics inhibit the reuptake, for reuse, of norepinephrine and serotonin already released in the brain. This means that greater amounts of these neurotransmitters are left in the brain, increasing their effects. Tricyclics also often affect the neurotransmitter acetylcholine (see Sir Henry Hallett Dale and Otto Loewi, p. 59), which causes a variety of side effects. SSRIs—including Prozac, Pfizer's Zoloft (sertraline), and SmithKline Beecham's Paxil (paroxetine)—work much like tricyclics, except that they are more selective; by affecting *only* serotonin, they cause fewer side effects.

The story of Prozac began in the 1960s, when Ray Fuller came to work at Eli Lilly. Fuller was born and raised in Illinois. He earned a bachelor's degree in chemistry in 1957 and a master's degree in microbiology in 1958, both from Southern Illinois University. Fuller worked his way through college as an employee at a state mental hospital; there he developed his interest in the brain and therapies for treating mental disorders. In 1961 he earned his Ph.D. in biochemistry from Purdue University. Shortly thereafter Fuller accepted a post as director of the biochemistry research laboratory at the Fort Wayne State Hospital for Mentally Retarded Children. In 1963 he joined Eli Lilly and Company as a pharmacologist and was set to work testing potential new antidepressants.

In his research Fuller had used rats treated with chloroamphetamine, which inhibited the production of serotonin, to measure the effects of other drugs on serotonin levels. Fuller believed that this method would forward research on brain chemistry and sought to entice Bryan Molloy, another Eli Lilly scientist, to this line of research.

Bryan Molloy was born in 1939 in Broughty Ferry, Scotland. He received his B.S. in chemistry in 1960 from the University of St. Andrews, and three years later earned his Ph.D. in chemistry from the same institution. In 1963 Molloy came to the United States to conduct postdoctoral research at Columbia and Stanford Universities. In 1966 he joined Eli Lilly as a senior organic chemist.

Molloy's research interests centered on the neurotransmitter acetylcholine and its effect on the heart. Fuller suggested that he abandon his cardiac work and take up a related line of research: looking for an antidepressant that did not affect acetylcholine. A new medication with that trait would not have many of the side effects that tricyclics had.

After accepting this proposal, Molloy began to research possible compounds for his project. Knowing that antihistamine research had led to some of the first antidepressants, including chlorpromazine (see Paul Charpentier et al., p. 118), Molloy developed compounds whose chemical structures closely resembled those of antihistamines.

David Wong, another Eli Lilly researcher with an interest in neurochemistry, joined the antidepressant project in 1971. Born in Hong Kong in 1935, Wong was the son of a machinist. As a youth, he was encouraged by his family to find a career helping people. Wong began his higher education at National Taiwan University, majoring in chemistry, and in 1957 he traveled to the United States to study at Seattle Pacific College, where he earned his B.S. in chemistry in 1960. In 1964 he earned an M.S. in biochemistry from Oregon State University. Two years later he received a Ph.D. in biochemistry from the University of Oregon. Wong joined Eli Lilly as a research biochemist in 1968.

In 1971 both Molloy and Wong attended a lecture on neurotransmission given by Solomon Snyder at Eli Lilly. A researcher from Johns Hopkins University, Snyder had developed a technique that would prove immensely useful to the Lilly team. He had ground up rat brains, separated out the nerve endings, and created an extract of nerve endings that worked in the same way as living nerve cells. Wong used this technique to test the effects of Molloy's compounds, one of which was found to block the reuptake of serotonin while affecting virtually nothing else. The compound was further tested in Fuller's chloroamphetamine-treated rat and was again found to block the reuptake of serotonin. This compound was fluoxetine.

Prozac was introduced in 1988, to great acclaim by doctors and patients alike. It is as effective as other antidepressants on the market, but it does not induce as many side effects. Furthermore, SSRIs are not lethal in overdoses. In the past decade Prozac has been shown to be effective in treating other conditions, including obsessive-compulsive disorder and bulimia nervosa.

# 13. Cancer
# and
# AIDS Drugs

**To fight the multitude of diseases that stem from the unbridled multiplication of the body's own cells, as in cancer, or of the cells of some infectious agent, researchers recognized in the 1940s that the cell's own genetic material could provide opportunities for drug development.**

Indeed, this line of research, which continues to this day, was undertaken before researchers solved the structure of the DNA molecule and understood genetic coding. Still, using whatever knowledge of the chemistry and mechanisms of genetics was available at the time, researchers synthesized substances that interfered with the reproduction of the cancer cells, viruses, or other cells without unduly damaging the body's normal cells.

Thanks to this approach, drugs have been developed to halt many diseases, including some like childhood leukemia or AIDS that in the recent past meant imminent death. In many respects the biotechnology revolution that is described in the following chapter forms a continuum with this long-standing strategy of intervening in the cell's own reproductive processes.

**Guanine.** In an early example of "rational design," George Hitchings and Gertrude Elion decided to interfere with the incorporation of guanine into genetic material as a way of inhibiting the overproduction of cells that characterizes cancer.

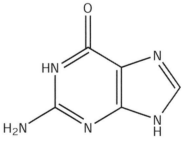

# George Hitchings (1905–1998) and Gertrude Elion (1918–1999)

Historically, drug developments frequently resulted from a trial-and-error process. As a consequence the element of chance has been essential in developing new pharmaceuticals. George Hitchings and Gertrude Elion diverged from this traditional path in their research, using what today is termed "rational drug design." They methodically investigated areas in which they could see cellular and molecular targets for the development of useful drugs. During their long collaboration Hitchings and Elion produced a number of effective drugs to treat a variety of illnesses, including leukemia, malaria, herpes, and gout.

George Hitchings, the son of a shipbuilder, was born in 1905 in Hoquiam, Washington. His family moved frequently along the West Coast during Hitchings's early years. When Hitchings was only twelve, his father died, an event that led Hitchings ultimately to pursue a career in medicine.

Hitchings demonstrated his early interest in science through his selection of courses at Franklin High School in Seattle. When he graduated from high school, his salutatorian address centered on the life and work of Louis Pasteur, whom Hitchings greatly admired for his commitment to basic research that led to practical developments.

Hitchings enrolled at the University of Washington in 1923, choosing to major in chemistry, primarily because of the enthusiasm of the chemistry faculty for their subject. He received his bachelor's degree in 1927 and stayed on at the University of Washington to earn a master's degree in 1928. Hitchings then attended Harvard University, where he received his Ph.D. in biochemistry in 1933. His work at Harvard centered on analytical methods used in physiological studies of purines, which are a class of compounds, including caffeine, adenine, and guanine, composed of a two-ringed structure containing carbon and nitrogen.

For the next decade Hitchings held a variety of temporary appointments at institutions like Western Reserve University (now Case Western Reserve) and the Harvard School of Public Health. In 1942, however, he joined the Wellcome Research Laboratories—then located in Tuckahoe, New York—as a biochemist. This research facility was operated by the British pharmaceutical firm, Burroughs Wellcome and Company, which is now part of GlaxoSmithKline. Two years later Hitchings hired Gertrude Elion as a labora-

tory assistant, thereby beginning a lifelong collaboration on drug development.

Gertrude Elion was born in 1918 in New York City, the daughter of a successful dentist. A shy young woman, Elion was an excellent student, and her parents encouraged her to attend college. The death of her grandfather from stomach cancer shortly before she began her studies at Hunter College prompted Elion to devote her life to medicine. This dedication was renewed when, soon after her graduation in 1937 with a degree in chemistry, her fiancé died of subacute bacterial endocarditis, an inflammation of the heart lining.

Elion knew that she needed to have a Ph.D. to do laboratory research. However, despite her excellent academic record, she was unable to get either a graduate fellowship or an assistantship, and so she began to look for a job. There, too, she had a great deal of difficulty, mostly owing to gender discrimination. At one point she enrolled in secretarial school to get some practical skills. After a series of temporary jobs Elion decided to volunteer her time in a chemistry laboratory; six months after she started, she was put on the payroll. Having saved enough money after a year and a half, Elion enrolled at New York University as a master's student in chemistry, working part-time as a doctor's receptionist and later as a substitute teacher to help pay the expenses. Nights and weekends were spent in the laboratory at the university doing research.

With the ongoing war, jobs in chemistry were beginning to open up for women. After earning her master's degree, Elion took a job testing food products for A&P grocery stores. She learned a lot about instrumentation but left when the position became too routine. She was then hired by Johnson & Johnson to synthesize sulfonamides, but the lab closed after six months. In 1944 she joined Wellcome Research Laboratories, intending to stay only as long as she continued to learn new things. When the NYU authorities informed her that it would be necessary to become a full-time student to complete a doctoral degree, Elion decided that she did not want to leave the exciting research at Wellcome to do this. She stayed there for the remainder of her career.

Rather than the traditional trial-and-error method of drug discovery, Hitchings believed in the necessity of a

■ Gertrude Elion and George Hitchings in the laboratory, 1948. In the background on the right is Elvira Falco, the first woman hired by Hitchings to work in his laboratory. Elion was the second. Courtesy GlaxoWellcome Inc. Heritage Center.

more rational method of research. The recent development of sulfa drugs (see Gerhard Domagk, p. 65) led him to think that other substances that interfered with the metabolism of microbes—as sulfa drugs had been shown to do—could also be developed as drugs. As a result Hitchings began examining the nucleic acids—now known as deoxyribonucleic acid (DNA) and ribonucleic acid (RNA)—and assigned Elion to investigate purines, including adenine and guanine, two of DNA's building blocks. They soon discovered that bacterial cells cannot produce nucleic acids—a material that determines the genetic makeup of a cell and directs the process of protein synthesis—without the presence of certain purines. They then set to work on antimetabolite compounds, which locked up enzymes necessary for incorporating these purines into nucleic acids.

By 1950 this line of research had paid off. Elion and Hitchings synthesized two substances—diaminopurine and thioguanine, which the enzymes apparently latched onto instead of adenine and guanine. These new substances proved to be effective treatments for leukemia. Leukemia is a form of cancer characterized by a great increase in the number of white blood cells in the body resulting from the activity of oncogenes, genetic material that has the ability to cause cancer. Elion later substituted an oxygen atom with a sulfur atom on a purine molecule, thereby creating 6-mercaptopurine (also known as 6-MP and trade named Purinethol), a molecule closely related to thioguanine. The new material, 6-MP, was also used to treat leukemia. But with all of these new treatments the disease was not cured; patients went into remission but then relapsed and died. Elion decided to examine everything about 6-MP, devoting six years of her life to this research. She discovered that treating childhood leukemia with a combination of 6-MP and one of several other drugs is more effective than using 6-MP alone. This method of treatment cures most patients.

After this success Elion and Hitchings developed a number of additional drugs by using the same principle that had led them to 6-MP. Another form of leukemia could be treated with 6-thioguanine. Later, these related drugs were found not only to interfere with the multiplication of white blood cells but also to suppress the immune system. This latter discovery led to a new drug, Imuran (azathioprine), and a new application, organ transplants. Imuran suppressed the immune system, which would otherwise reject newly transplanted organs. The team also developed allopurinol (Zyloprim), a drug that reduces the body's production of uric acid and therefore can be used for treating gout, a painful condition caused by the buildup of uric acid in the joints.

In the 1960s Elion and Hitchings shifted their research to nucleic acid formation in lower animals and the differences between these processes in animals and in people. They determined that infectious diseases could be fought if drugs could be targeted to attack bacterial and viral DNA. This work resulted in pyramethamine—used to treat malaria—and trimethoprim (Septra)—used to treat meningitis, septicemia, and bacterial infections of the urinary and respiratory tracts.

As a result of these overwhelming successes Hitchings was promoted in 1967 to vice president of research; this promotion ended his active participation in research. Elion, too, was promoted—to head of the department of experimental therapy. Despite her new responsibilities Elion continued her research and was essential in the development of acyclovir, an antiviral drug effective against herpesvirus. Although the drug was originally synthesized by Howard Schaeffer, Elion determined exactly how and why it worked. Acyclovir, marketed as Zovirax, interferes with the replication process of the herpesvirus—and only the herpesvirus—proving that drugs can be selective. This principle led to the eventual development of AZT (azidothymidine) by Elion's colleagues. AZT is used to treat AIDS.

In 1970 the Wellcome Laboratory moved to Research Triangle Park, North Carolina, and Elion and Hitchings moved along with it. Elion retired in 1983, eight years after Hitchings. Along with James Whyte Black (p. 148), they received the 1988 Nobel Prize in physiology or medicine.

# IRVING SIGAL (1953–1988), EMILIO EMINI (1953– ), JOEL HUFF (1946– ), JOSEPH VACCA (1955– ), BRUCE DORSEY (1960– ), AND JON CONDRA (1949– )

In recent times the acquired immunodeficiency syndrome (AIDS), first reported by medical scientists in 1981, has gripped the attention of the public as an awful killer of young adults supposedly in the prime of their lives. Told here is the story of just one of the drugs developed to combat AIDS, Merck's Crixivan. Other companies have succeeded in the race to find effective treatments. Crixivan was the third drug with its kind of action on the human immunodeficiency virus (HIV), the causal agent in AIDS, to be approved by the U.S. Food and Drug Administration (FDA); Invirase, from Hoffmann–La Roche, and Norvir, from Abbott Laboratories, were approved earlier.

Merck began searching for a means to control AIDS in 1986, when many of the basic characteristics of the virus that causes the disease were still unknown. Researchers, however, already understood that the virus's genetic material consists of ribonucleic acid (RNA), which it can transcribe into deoxyribonucleic acid (DNA). The virus then injects this DNA into host cells, where it is incorporated into the host's DNA. This makes the host cells, usually the helper T-cells vital to the immune system, produce new AIDS viruses at an incredible rate—in the process killing the host cells and leaving the victim's body open to bacterial and viral invaders of all kinds.

The Merck scientists at this time included Irving Sigal, a biochemist who had received his Ph.D. from Harvard University in 1978 and had been with the company fewer than five years. Emilio Emini soon joined forces with Sigal as co-champion of Merck's AIDS research. He was hired in 1983, shortly after completing his doctoral degree in microbiology from Cornell University. When they and other Merck scientists began their AIDS research, azidothymidine (AZT), developed by Burroughs Wellcome scientists (see George Hitchings and Gertrude Elion, p. 170), was just about to receive approval from the FDA. AZT is a reverse transcriptase inhibitor: that is, it competes with the enzyme reverse transcriptase for sites on the DNA chain being formed, thus preventing transcriptase from enabling the further buildup of the chain by polymerization. In short, AZT prevents HIV's replication. Although the discovery of AZT (and other reverse transcriptase inhibitors) was considered a remarkable breakthrough in combating AIDS, there were problems with side effects, and even more important, the virus was able to mutate easily to become resistant to the drug.

Merck researchers pursued several parallel approaches to finding an AIDS drug, with different levels of intensity at various times. Initially, it seemed that a vaccine could be developed. Using attenuated, or weakened, viruses to stimulate an immune response capable of resisting future infection appeared too risky, but perhaps a subunit common to many of HIV's surface proteins could be found that would evoke an immune response—a tactic that had just been successfully used in Merck's bioengineered vaccine for hepatitis B (see William J. Rutter, p. 193). This line of attack, led by Emini, continued at Merck until 1993, when the researchers concluded that there were just too many varieties of surface protein to mimic to produce an effective vaccine.

■ Irving Sigal, 1988. Courtesy Merck Archives, Merck & Co., Inc.

■ Jon Condra (left) and Emilio Emini, early 1990s. Courtesy Merck Archives, Merck & Co., Inc.

Research on other reverse transcriptase inhibitors offered another pathway. Burroughs Wellcome's researchers had chosen AZT because of its structural similarity to the components of the nucleic acids RNA and DNA, but non-nucleoside analogues might make it more difficult for the virus to adapt to the presence of the new drug. Sigal had become convinced early on that an alternative way to stop HIV in its tracks lay in the inhibition of another enzyme—a protease that catalyzes a critical step near the end of the HIV replication cycle.

In 1987 a group in medicinal chemistry, led by Joel Huff, started the search for a protease inhibitor by screening hundreds of compounds, many of which they designed themselves. Huff had worked at Merck since 1974, when he received his doctorate in organic chemistry from MIT. Working in Huff's group was Joseph Vacca, who joined Merck in 1981, just after receiving his Ph.D. from the State University of New York at Buffalo.

This new project was helped immeasurably by the Merck scientists' experience with looking for an inhibitor of the angiotensin-converting enzyme (ACE), which is also a protease. Their research paralleled Miguel Ondetti's work at Squibb (see Ondetti, p. 151) and resulted in Merck's Vasotec. And more was known about the protease they were targeting this time. It had been copied a thousandfold by biotechnology techniques (see Paul Berg et al., p. 179) so that it could be fully characterized—including its molecular structure. But in December 1988 the distressing news arrived that Sigal had been killed in the crash of Pan Am 103 over Lockerbie, Scotland.

Still the Merck researchers soldiered on, and in early 1990 trials were begun to determine the safety and potency of selected antiviral product candidates. Candidate L-689,502, or simply 502, a protease inhibitor, seemed a likely prospect, although it was not orally active and would have to be injected. When it proved toxic, the protease inhibitor line of research was put on the back burner and the reverse transcriptase inhibitor line was taken up in earnest. But by 1991 it was clear that the virus could easily develop resistance even against non-nucleoside analogs to reverse transcriptase.

Interest once again focused on protease inhibitors. Huff and Vacca tried to piece together a new protease inhibitor from the best parts of 502 and some elements borrowed from the protease inhibitor on which scientists at Hoffmann–La Roche were working. They did not succeed in their effort; so by the end of 1991 they decided to create a new team to work on the problem—one headed by Bruce Dorsey.

Dorsey had just completed his Ph.D. at the University of Pennsylvania in 1987 and had only been at Merck since 1989. Dorsey proceeded to change the basic ring system on the left-hand side of the molecule that Huff and Vacca had been working on. This new compound, L-735,524, known as 524, inhibited the targeted protease, retained activity when administered orally to animals, and passed toxicity tests. It was the most complicated drug candidate ever synthesized by Merck scientists. Its components can be arranged in three-dimensional space in thirty-two possible configurations, only one of which possesses the requisite activity and tolerability.

■ Joel Huff, 1989. Courtesy Merck Archives, Merck & Co., Inc.

■ Joseph Vacca, 1982. Courtesy Merck Archives, Merck & Co., Inc.

■ Bruce Dorsey, 1989. Courtesy Merck Archives, Merck & Co., Inc.

Compound 524 was later named indinavir sulfate, or Crixivan. However, scientists thought it might be possible for HIV to mutate around this compound, much like it did around AZT, thereby rendering the drug candidate no better than the AIDS drugs already on the market. This problem was tackled by Jon Condra, a biologist trained at the University of California at Riverside, from which he received his Ph.D. in 1979. He joined Merck four years later. Condra's approach to the problem was to try to determine what HIV would have to do in order to resist Crixivan. After much work he was able to predict that HIV could develop a resistance to the new drug—a prediction that was fulfilled.

In early 1993 the FDA granted Merck's request to move compound 524 into human trials. At first these trials seemed to go very well, resulting in a substantial decline in HIV levels in the study subjects. But a year later viral resistance to the drug appeared to have developed in most of the early patients. The clinicians then increased the amount of 524 in the dosage and gave it in combination with AZT to the patients in the clinical trials. They reasoned that attacking two sites in HIV's replication cycle might meet with greater success in keeping the virus at bay than focusing on a single site; this theory proved correct in practice.

Meanwhile an internal decision was made at Merck to prepare full-scale production facilities so that once FDA approval was gained, the company would be prepared to meet the demand. Starting in the spring of 1995, staff members of the manufacturing division dedicated themselves to meeting a deadline that was moved ever closer. On all sides—at Merck, at the FDA, and among AIDS activists—there was pressure for accelerated approval of Crixivan. The actual submission was made on 31 January 1996, and on 13 March 1997 the FDA approved Crixivan for use in adults.

177

# 14. Biotechnology

The biotechnology industry has revolutionized the kinds of pharmaceuticals that are available and how they are produced. The technologies that this industry deploys depend on knowledge of the molecular structures of the substances of life—gained over decades of biochemical research—and of the molecular details of reproduction—knowledge gained through the relatively new science of molecular biology.

After the discovery of the first antibiotics, scientists relied on microorganisms to produce substances like penicillin, which are the microbes' own natural products commandeered to fight human diseases. Now genetic engineering allows scientists to make microbes and tumor cells produce something other than their own natural substances—substances that are natural to the human body but in critically short supply. Through recombinant DNA processes and hybridoma technology (described in the succeeding biographical essays), scientists have succeeded in synthetically producing hormones like insulin and erythropoietin as well as a new hepatitis vaccine, monoclonal antibodies used to diagnose and treat a number of diseases, and interferon, a natural virus fighter. And the list of such pharmaceuticals continues to grow. Biotechnology also promises to bring about the correction of errors in the human genome itself, a topic discussed briefly in the final chapter of this book.

The first stages of the commercial development of biotechnology saw the founding of many small companies funded by venture capital in which university scientists participated as inventor-entrepreneurs. The world had not previously witnessed the birth of so many companies in the area of pharmaceuticals in such a short time. Most of these new companies—if they succeeded in their research objectives—were not prepared to scale up processes, conduct clinical trials, or manufacture a product. Established pharmaceutical companies entered rather quickly into biotechnology in a variety of ways, including joint ventures, and these companies began to develop their own biotech research capabilities.

■ Paul Berg opening a jar under a protective hood. Courtesy Stanford University Archives.

## PAUL BERG (1926– ), HERBERT W. BOYER (1936– ), AND STANLEY N. COHEN (1935– )

The invention of recombinant DNA technology, the way in which foreign genetic material is artificially introduced into the genome of another organism and then replicated and expressed by that other organism, was largely the work of three men—Paul Berg, Herbert Boyer, and Stanley Cohen—although many other scientists made important contributions to the new technology as well.

Berg grew up in Brooklyn, New York, in the 1930s. His interest in science was stimulated by his reading of Paul De Kruif's *Microbe Hunters* (1926) and Sinclair Lewis's *Arrowsmith* (1925). After graduating from high school at the age

of sixteen, Berg had some difficulty deciding what and where he should study. Perusal of a catalog from Pennsylvania State College (now University) alerted him to the existence of the field of biochemistry, and he was soon on his way to Penn State. In the middle of his college years his enlistment in the U.S. Navy was activated, but World War II ended before he saw combat. He returned to Penn State to complete his degree and then went on to Western Reserve University (now Case Western Reserve) to get a Ph.D. in biochemistry. He pursued postdoctoral research on enzymes first with Herman Kalckar in Copenhagen and then Arthur

179

Kornberg at Washington University in St. Louis. In 1955 Berg was appointed to the faculty at Washington University where he became a leader in deciphering the biosynthesis of proteins on the basis of codes carried on deoxyribonucleic acid (DNA) and ribonucleic acid (RNA) molecules.

In 1959 Berg joined the faculty of Stanford University. There he became interested in the genetics of microbes and took a leave to study at Renato Dulbecco's laboratory at the Salk Institute, where he learned the techniques of animal-cell culture. Dulbecco had already shown that certain viruses induce a cancerous state in an infected cell by taking over the expression of the genetic information of that cell

for their own reproduction. Like other scientists at the time, Berg began to wonder whether it would be possible to insert foreign genes into a virus, thereby causing it to become the vector by which genes could be carried into new cells.

Berg's 1971 landmark gene-splicing experiment (Figure 1) involved splicing a bit of the DNA of the bacterial virus known as lambda, one of the "phages" that invade bacteria, into the DNA of simian virus SV40, whose natural host is the monkey. The DNA of both these viruses occurs in closed loops. In the first step of Berg's experiment the loops were each cut in one place by an enzyme, Eco RI. This enzyme had just been discovered in Herbert Boyer's

■ Herbert Boyer. Courtesy Albert and Mary Lasker Foundation.

laboratory at the University of California at San Francisco (UCSF). It was found in a plasmid, DNA that occurs mostly in bacteria and is physically separate from the DNA making up a given bacterium's chromosomes—in this case, in *Escherichia coli*, a common bacteria found in the human digestive tract, among other places. Next, to make the ends of these now-linear molecules stick together again, they were modified by two other enzymes, using a procedure developed by Peter Lobhan, a graduate student in the laboratory of Berg's Stanford colleague Dale Kaiser. Then the two types of DNA were mixed together where they rejoined into loops in such a way that the new loops combined DNA from each source. Berg's gene-splicing experiment resulted in the first man-made recombinant DNA (rDNA), as such molecules came to be called. The award ceremony for Berg's 1980 Nobel Prize in chemistry, shared with Walter Gilbert and Frederick Sanger, highlighted this work.

Berg did not immediately take the step of introducing the rDNA into another organism because of the public controversy over the potential dangers of such experimentation. The fear at the time was that rDNA carrying a dreaded gene—for example, for the creation of cancerous tumors—might escape the laboratory in some common bacteria and be spread everywhere. As chair of the National Academy of Science's Committee on Recombinant DNA Molecules, Berg played an active role in the debate among scientists and with the public about potential limitations on such research. In the 1970s the National Institutes of Health, which had become the chief financial supporter of this and most other types of biomedical research, issued guidelines for the safe conduct of rDNA research. Over time these guidelines have been eased, as more experience has shown the hazards to be far less than imagined.

The next landmark in the development of modern biotechnology was the insertion of rDNA into bacteria in such a way that the foreign DNA would replicate naturally (Figure 2). This step was taken in 1972 by Boyer at UCSF, in collaboration with Stanley Cohen of Stanford University.

Boyer was born and raised in western Pennsylvania and attended St. Vincent's College in Latrobe. There he enrolled in premedical studies. However, he soon was captivated by research and chose to major in chemistry and biology. He completed a Ph.D. in biochemistry at the Uni-

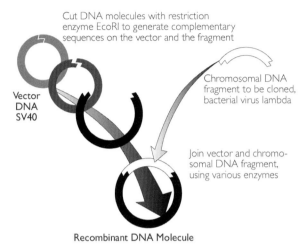

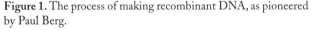

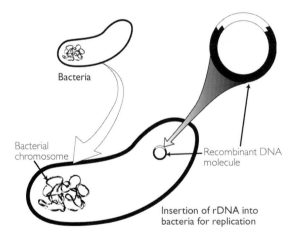

**Figure 1.** The process of making recombinant DNA, as pioneered by Paul Berg.

**Figure 2.** The insertion of recombinant DNA so that the foreign DNA will replicate naturally, as pioneered by Herbert Boyer and Stanley Cohen.

versity of Pittsburgh and then went on to a postdoctoral fellowship in the laboratory of Sidney Alberg at Yale University. In 1966 he accepted an appointment at UCSF, which was becoming a center of excellence in the several disciplines that contributed to the emerging field of biotechnology (see William J. Rutter, p. 193).

In 1972 researchers, including Boyer, realized that the enzyme Eco RI cut DNA in such a way that the ends were not blunt but staggered, so that no molecular additions were needed to make one severed piece latch on to another piece possessing complementary cuts. Boyer and Robert Helling, a faculty member spending his leave from the University of Michigan at UCSF, began their effort to create rDNA to insert in *E. coli* by trying to use Eco RI to open up the DNA of bacterial virus lambda. They became frustrated, however, when the enzyme cut the DNA in five places instead of one, as desired.

November 1972 found both Boyer and Cohen in Hawaii giving papers at a United States–Japan joint meeting on plas-

mids. While Boyer was describing his data showing the nature of the DNA ends generated by Eco RI cleavage, Cohen was reporting on a procedure recently discovered in his laboratory that enabled bacteria to take up plasmid DNA and produce offspring that contained self-replicating plasmids identical to the original implant—clones. Over sandwiches late one night at the conference, the two men laid plans for a collaborative project to discover what genes are present on plasmids and how they are arranged.

A native of Perth Amboy, New Jersey, Cohen received his undergraduate education at Rutgers University and then proceeded to the University of Pennsylvania for an M.D. After completing his medical education, he began a full-time career in medical research and teaching at the Albert Einstein College of Medicine in New York. There he worked on the complex mechanisms that control gene expression in the bacterial virus lambda. In 1968 he accepted an appointment at the Stanford University School of Medicine.

The collaboration between Boyer and Cohen was very close. Plasmids isolated at Stanford were transported to Boyer's lab in San Francisco for cutting by Eco RI and for analysis of the DNA fragments. These were transported back to Stanford, where they were joined and introduced into *E. coli*, where they multiplied. Then the brand-new recombinant plasmids were isolated and analyzed in each laboratory. The transportation of the precious DNA samples became the responsibility of Annie Chang, a research technician who worked in Cohen's laboratory but lived in San Francisco.

The first success of the Boyer-Cohen collaboration occurred in spring 1973 and involved one of Cohen's plasmids, pSC101. Plasmids were already known to transfer drug resistance among bacteria, and this one could make *E. coli* resistant to the antibiotic tetracycline. The plasmid pSC101 was cleaved by Eco RI at only one site, leaving intact the plasmid's ability to replicate. When the linearized pSC101 DNA was mixed with other DNA that had been cleaved by the same enzyme, the complementary ends of fragments from both sources of DNA joined together into new loops. Treatment with another enzyme closed the still-visible nicks in the DNA loops, which were then introduced into calcium chloride–treated bacteria. The bacteria were spread on a culture containing tetracycline, and only the bacteria with the rDNA plasmids survived.

Boyer and Cohen soon moved on to more complicated cloning experiments. They joined tetracycline-resistant plasmids with kanamycin-resistant plasmids and inserted them in *E. coli*. Next they showed that genetic materials could indeed be transferred between species, thereby disproving a long-held myth. They snipped a piece of staphylococcus plasmid and spliced it with one of the many *E. coli* plasmids and inserted the whole in *E. coli*. The DNA from staphylococcus, a different species of bacteria, was successfully propagated in *E. coli*. An even-greater triumph of interspecies cloning was the insertion into *E. coli* of genes taken from the South African clawed frog.

Although they had not originally entered the field with this objective in mind, Boyer and Cohen and other scientists involved in cloning experimentation soon recognized the feasibility of using bacteria into which human genetic information was incorporated to duplicate the body's natural means of fighting disease and to remedy birth disorders. Alerted to this possibility by an article in the *New York Times*, the Stanford Office of Technology Licensing convinced Cohen and Boyer that a patent would serve to propagate recombinant technology more quickly and would generate funds that could be used in their respective universities. Formal application for the patent was made in fall 1974, and after extensive controversy the patent was granted in 1980.

Meanwhile, commercial ventures had started up, with the objective of capitalizing on the new technology. At the forefront was Genentech, a corporation founded in 1976 by Boyer and Robert Swanson, a young venture capitalist. Whereas Boyer became deeply involved in commercial work, Cohen remains an academic researcher, focusing on basic questions in genetics and biology. Over the years Cohen has consulted, though, for several biotechnology companies, as did Berg, who later helped found a company called DNAX.

In fall 1977, even before Genentech had its own facilities, Boyer at UCSF and Keiichi Itakura at the City of Hope Medical Center in Duarte, California, succeeded in expressing a mammalian protein in bacteria—somatostatin. This hormone, produced in the human brain, plays a major role in regulating the growth hormone. Recombinant somatostatin was shown to be virtually identical to the naturally occurring substance. In 1978 Boyer and Itakura also constructed a plasmid that coded for human insulin. By then they had many rivals, some of them small start-ups backed by large pharmaceutical companies. In the case of recombinant insulin Eli Lilly and Company signed a joint-venture agreement with Genentech to develop the production process for Humulin, thus playing a role similar to that taken by Lilly in 1922 with the University of Toronto scientists who had originally discovered insulin therapy (see Frederick Grant Banting et al., p. 52). In 1982 Humulin was approved by the FDA, and it became the first biotechnology product to appear on the market.

■ César Milstein examining a plate. Courtesy Medical Research Council Centre.

## CÉSAR MILSTEIN (1927– ) AND GEORGES KÖHLER (1946–1995)

The development in the 1970s of what is now known as hybridoma technology has had far-reaching effects, not only for the developers, César Milstein and Georges Köhler, who received a Nobel Prize in physiology or medicine for their work, but also for the biotechnology industry, which grew partially out of this technology. Monoclonal antibodies have various uses in today's medical world, for example, in transplantation and the treatment and diagnosis of a number of diseases, among them rheumatoid arthritis, cancer, and AIDS. Entire biotech companies have been founded on monoclonal antibodies, and the development of this technology has stimulated the growth of the biotech field more generally.

César Milstein was born in Bahia Bianca, Argentina. In 1952 he received his degree in chemistry from the University of Buenos Aires. Although there were few professional opportunities for scientists in Argentina, Milstein decided to remain at the University of Buenos Aires to obtain a doctorate in biochemistry, which he received in 1957. He then accepted a fellowship to work for the Medical Research Council in Cambridge, England, researching enzymes. For this work he received a Ph.D. from Cambridge University in 1960. Three years later political uncertainty in Argentina led Milstein to return to Cambridge, where he stayed for the remainder of his career.

When Milstein resumed his research at Cambridge, he began studying antibodies. Also known as immunoglobulins, these protective molecules are part of the body's immune system. By the mid-twentieth century the basic mechanism of antibody production was fairly well established. Antibodies circulate in the blood and lymph systems in response to the presence of antigens—foreign particles or microbes, including bacteria and viruses. Antibodies are produced in the lymphoid tissues by B-cells, which are a type of lymphocyte, or white blood cell, capable of recognizing antigens through surface receptor molecules. These receptor molecules bind to the surface of the antigens, stimulating the B-cells to multiply and produce massive quantities of antibodies that then attack the antigens. Each receptor can bind to only one specific antigen, thereby providing immunity against only that one antigen. However, more than one type of antibody can react with one antigen. For example, when an individual is infected with a virus, his or her body produces several different antibodies to combat that virus.

Milstein's research interests centered on differences in antibody specificity. Important to this type of research are myelomas. These malignant tumors in bone marrow produce B-cells that in turn produce astonishingly large amounts of a single antibody. Since such cells multiply indefinitely, they are easy to cultivate, unlike cells produced by normal tissue, which die after a certain number of divisions. As part of their investigation of the production and structure of antibodies, Milstein and his collaborators succeeded in genetically fusing two kinds of myeloma cells, from rat and mouse tumors, thereby forming cells with large nuclei containing the genome of both kinds of parent cells. They anticipated that these fused cells would produce thousands of antibodies for them to study—a phenomenon observed in normal antibody-producing cells, that each cell produces only one kind of antibody even though it has genetic code to produce other types. The cells they were trying to fuse were placed in a medium (hypoxanthine aminopterin thymidine, or HAT) that could separate the few fused cells from the vast majority of unfused cells; further, the cells were genetically marked so that fused cells could be easily located. An inactivated Sendai virus was used as the fusing agent. The resulting hybrids, contrary to expectation, continued to produce antibodies of both parental types. Georges Köhler learned of the success of the fusion process as applied to two antibody-producing sets of cells. He thought it might help him overcome his difficulty in producing enough antibodies from tissue culture of normal cells to continue the studies he had just begun.

Köhler was born in 1946 in Munich, Germany, to a German father and a French mother. In 1971 he received a degree in biology from the University of Freiburg. He continued his studies there, focusing on immunology, and conducted his research at the Institute of Immunology in Basel. Though his work was on enzymes, Köhler began to develop an interest in antibodies. When Milstein visited the institute in 1973, he invited Köhler to come and work with him at Cambridge. Thus, in 1974, when Köhler completed his Ph.D., he joined Milstein's laboratory.

Köhler had the idea of fusing myeloma cells with normal lymphocytes from the spleen of a mouse immunized to a specific antigen (sheep red blood cells) in order to create hybrid cells that could produce antibodies specific to this antigen (later known as monoclonal antibodies) in limitless amounts. He began with a line of myeloma cells that had specificity for a particular antigen, but he was unable to grow it. He therefore developed a mutant of the myeloma cells and fused it with the normal lymphocytes. He then combined his hybrids with his selected antigen and was pleased to see them react. The process of producing such hybrids is now known as hybridoma technology.

In the midst of his research Köhler returned to Basel. He tried a number of variations on his cultures, and although his first experiments were unsuccessful, he soon identified the problem: a toxic batch of a reagent. He eventually refined the hybridoma technology further, fusing the lymphocytes with a strain of myelomas that no longer secreted antibodies. This procedure made selecting the type of antibody to be produced much simpler.

Milstein and Köhler did not patent their technique. They felt their discovery was basic science and should be available to all. A bitter dispute raged between British and American researchers when, in 1979, a team of scientists at the Wistar Institute in Philadelphia patented hybridoma technology as it applied to tumor and viral antigens.

Monoclonal antibodies have a variety of medical applications. When tagged with radioisotopes, these antibodies can show the location of tumors in the body, thereby aiding in the diagnosis of different forms of cancer. Monoclonal antibodies can also distribute drugs to specific sites in the body and thus combat a number of diseases. They can also be used to counter the body's own immune system, which makes the body naturally reject organ transplants and, when it is overactive, causes arthritic conditions.

■ Georges Köhler. Courtesy Medical Research Council Centre.

# George Rathmann (1927– ) and Fu-Kuen Lin (1941– )

The story of recombinant human erythropoietin shows yet another picture of how entrepreneurial and scientific intelligence combined to produce one of the great successes of biotechnology. Erythropoietin, a hormone manufactured in tiny quantities in the kidneys, travels through the bloodstream to the bone marrow, where it stimulates the production of red blood cells, or erythrocytes. When the kidneys fail to produce sufficient erythropoietin, debilitating anemia results. Before the development of recombinant erythropoietin, patients suffering from chronic kidney disease and renal failure had to undergo not only dialysis to filter waste products from their blood but also frequent blood transfusions to correct anemia. The work of George Rathmann, chief executive officer of the biotech company Amgen, and Fu-Kuen Lin, a researcher at Amgen, spared hundreds of thousands of anemic patients from expensive treatments that carried the risk for hepatitis and AIDS infections.

George Rathmann grew up in Milwaukee, Wisconsin, and graduated from Northwestern University with a bachelor's degree in physical chemistry. He went on to get his Ph.D. in that field from Princeton University in 1952. Returning to the Midwest to start his career, Rathmann took a position with the 3M Company in St. Paul, Minnesota, where he continued to pursue research he had begun in graduate school: the scattering of light by polymers.

Rathmann moved quickly through the ranks at 3M, from research to managerial positions. After two decades with the company Rathmann went on to join the Diagnostics Division of Abbott Laboratories, a well-established pharmaceutical company in Chicago.

While at Abbott, Rathmann worked on diagnostics and vaccines for hepatitis B, which at that time relied on a weakened form of the virus. He realized that a synthetic antigen—one created through genetic engineering—had great potential as a substitute for the weakened virus. In the late 1970s he contacted Winston Salser, a professor in the Molecular Biology Institute at the University of California at Los Angeles, for a clone of the hepatitis B antigen—the substance on the surface of the virus that stimulates the body's production of antibodies to the virus. From this clone, Rathmann reasoned, his recombinant DNA research group

at Abbott could begin the work involved in making a synthetic surface antigen (see William J. Rutter, p. 193).

This early biotechnology project never got under way at Abbott, but Rathmann's initial approach to UCLA led him to the expanding world of biotech. Winston Salser was the founding member of the scientific advisory board behind a nascent Southern California biotech company, Applied Molecular Genetics, or Amgen. Five major biotech companies already existed at the time, but venture capitalists and university scientists interested in launching Amgen met in the spring of 1980 and decided that the field could support another start-up. Salser remembered George Rathmann and asked whether he was still interested in recombinant DNA. Shortly after Amgen's incorporation in April 1980, Rathmann was offered the position of president and CEO at the new company.

Amgen's first successful drug research project came through one of its scientific consultants, Eugene Goldwasser, a professor at the University of Chicago who had isolated a small amount of the human hormone erythropoietin. The existence of erythropoietin was first recognized in 1906 by researchers at the University of Paris. These scientists discovered a substance in the serum of animals that had been recently bled to make them anemic. They called it hemopoietine because it provoked an overabundance of red blood cells when introduced into healthy animals without anemia. It was not until 1960, however, that hemopoietine, which had been renamed erythropoietin, was isolated from the blood plasma of anemic sheep, which produced the substance in quantity in order to trigger the production of the needed red blood cells. Almost two decades later human erythropoietin was isolated from the urine of anemic patients. Researchers at Amgen, who had a small amount of pure erythropoietin from Goldwasser's lab, set out to find the genetic code for the hormone—that is, the information necessary for synthesizing it. With a blueprint of the genetic code, producing a genetically engineered version would be possible.

Fu-Kuen Lin was hired by Amgen in August 1981, after the company secured enough financing to launch their research venture. Born and educated in Taiwan, Lin came to

■ George Rathmann surrounded by fermentation tanks for growing microorganisms. Courtesy Amgen, Inc.

■ Fu-Kuen Lin conducting research at his desktop computer. Courtesy Amgen, Inc.

the United States in 1967 to pursue a doctoral degree in plant pathology at the University of Illinois at Urbana-Champaign. He then held a number of positions in academia in Taiwan and the United States, including a postdoctoral fellowship at the Medical University of South Carolina, where he became familiar with recombinant DNA techniques.

After approximately two years of grueling, meticulous work, Lin and his colleagues found the human gene responsible for the production of erythropoietin. This feat has been likened to finding a needle in a whole field of haystacks. Human chromosomes contain some hundred thousand genes, each made up of about three to five thousand nucleotides, the four small building blocks (adenine, thymine, cytosine, and guanine) that make up DNA and—in combinations of any three—code for specific amino acids. Even with Lin's strategies for homing in on the objective, which were critical to the success of the project, there remained thousands of experiments to conduct. Some scientists at Amgen came to view the search as an impossible mission, although it was known that other biotechnology companies were seeking the same objective.

Fortunately for the success of the project, Lin was extremely persistent. Through it all George Rathmann remained supportive—only once threatening to end the project. When more resources were needed, he allocated them. Beginning with a workforce of only two or three people, the project grew to engage the labors of twenty or so people—about half the total number of workers at Amgen at that time—by the end of 1983, when they were isolating the human gene.

The project began with the partial identification of the linear sequence of amino acids in the tiny sample of pure erythropoietin supplied by Goldwasser's lab. Lin led the project, designing and synthesizing oligonucleotides, short segments of DNA about seventeen to twenty nucleotides in length, based on the genetic codes for these amino acids.

Large mixtures of these oligonucleotides were radioactively tagged and used as probes to search for potential matches for the gene responsible for producing erythropoietin. To prepare for this matching process, pieces of human DNA that might contain the gene for erythropoietin were cloned into a bacteriophage, a virus that normally infects bacteria, to produce large quantities of DNA. The pieces of human DNA were retrieved, dissociated into their two strands using an alkaline solution, and affixed to positively charged nylon sheets. Researchers then added the radioactively tagged probes, which would form complexes only when the probe's sequence "matched" (i.e., complemented) the sequence of the human erythropoietin gene. The next step was to expose X-ray film to the nylon sheets. Lin and his colleagues could then locate the matches by noting where the dark spots were on the film—indicating that the radioactive probe had attached there. Once the gene was located in this manner, it could be isolated, sequenced, and analyzed further. The human erythropoietin gene was subsequently cloned and, so that it would produce erythropoietin itself, inserted into mammalian cells—Chinese hamster ovarian cells, which possess a very stable chromosome content in culture.

Other scientists at Amgen handled the first scaling up of genetically engineered erythropoietin, designing the processes by which enough hormone could be cultured so that tests for safety and efficacy could be run. Clinical trials were begun in December 1985, and FDA approval was granted in June 1989. Since 1989 the uses of Amgen's genetically engineered erythropoietin, called Epogen, have widened to include treatment of the anemias caused by the HIV-fighting drug azidothymidine (AZT) and by cancer chemotherapy. Epogen can also be administered to individuals scheduled to undergo an operation who want to use their own banked blood instead of transfusions from other sources. By speeding up the formation of red blood cells, Epogen shortens the time needed to bank enough blood.

# Charles Weissmann (1931– )

The development of recombinant DNA techniques created a revolution in science. It enabled scientists to design projects that had previously been unthinkable. The cloning of interferon is one example of this phenomenon. In 1978 Charles Weissmann began collaborating with Kari Cantell, with the goal of cloning the interferon gene. Neither Weissmann nor Cantell wholly believed they would succeed, yet because of their expertise and the technology available, they did.

Interferon is a protein that occurs in various forms and is secreted by cells in response to viral infection. It in turn stimulates the production of other proteins that prevent the virus from replicating. There are three major forms of interferon—alpha, beta, and gamma—and even more forms are known to exist.

The history of interferon began in 1937, when two British scientists first described "interference," a physiological phenomenon that created a natural shield against viral infection. Nearly twenty years later Jean Lindenmann, a Swiss biologist, and Scottish-born Alick Isaacs pioneered research on interferon, together identifying the substance in 1956–57. Through their work they discovered that interferon was a protein. Lindenmann and Isaacs published their research, but their results were considered highly doubtful by a large portion of the scientific community. Research continued, however, and interferon came to be known as "antiviral penicillin," that is, a wonder drug able to combat viral diseases. The Scientific Committee on Interferon was formed in 1959, with Isaacs as chair. Clinical trials showed some promise, but by the mid-1960s excitement about interferon had died down.

In 1963 the Finnish virologist Kari Cantell decided to study the medical potential of interferon. Two years later, at the State Serum Institute in Helsinki, Cantell began collaborating with Hans Strander. This scientific duo tested interferon on humans who had various types of cancer that were thought to have viral origins. Cantell became the single producer of clinical-grade interferon in the world. He produced approximately one gram of interferon per year, using leukocytes, or white blood cells, from ninety thousand blood donations to the Finnish Red Cross. Because cells produce minuscule amounts of the protein, there was a severe shortage of interferon for testing. Scientists needed a synthetic method for producing interferon.

Charles Weissmann was professor of molecular biology at the University of Zurich when he approached Cantell with the project of cloning interferon. Weissman was born in Switzerland in 1931, where his father was a famous film producer and distributor. He spent most of his childhood in Brazil, though, where his family lived during World War II. Weissmann received his diploma from the University of Zurich and then spent seven years at New York University, where he studied with Nobel laureate Severo Ochoa, a man famous for his work on the biological synthesis of RNA and DNA.

Weissmann wanted to isolate, and then clone, the gene for leukocyte interferon, a procedure that would enable scientists to make unlimited quantities of the protein. The leukocytes that Cantell possessed contained unusually large levels of the messenger RNA (mRNA) for the interferon gene. The mRNA is a copy of the genetic information (DNA) from a gene (in this case, the gene for interferon), and mRNA transports the message from the nucleus to the cytoplasm, where the ribosomes produce the desired protein. A supply of these mRNA-rich cells, combined with Weissmann's knowledge of the latest recombinant DNA techniques, provided the scientists with a solid base for their research.

The goal was to produce clones of the gene for human interferon, and the research was supported by Biogen, a biotechnology company founded in 1978, with which Weissmann was intimately involved. Knowing that viral infection triggers interferon production, the collaborators set to work determining the relative interferon mRNA levels of leukocytes at different times after viral infection. By doing so, they hoped to find the time of peak production of mRNA.

Not long after the group began this work, Tada Taniguchi, an important member of Weissmann's laboratory, left the project to return to Japan. His departure added to the already fierce competition to clone interferon, as he continued his cloning work in Japan on a different form of the protein—fibroblast interferon, which is produced by fibroblast cells present in connective tissue. Throughout the interferon work, rumors abounded that Genentech, Cetus, and various other academic scientists and biotech companies had successfully cloned interferon.

■ Charles Weissmann. Courtesy C. Weissmann.

Adding to the already difficult situation, by 1979 Biogen had depleted its capital, forcing it to seek assistance from Schering-Plough Corporation. Schering-Plough agreed to invest in Biogen and in return obtained the rights to leukocyte interferon.

The strategy of the scientists was to transform into DNA all of the mRNA from the special leukocytes that Cantell had identified in order to locate the interferon gene and then clone it. This is the reverse of what naturally happens in the body. Typically, the DNA is transcribed into mRNA, which carries the message to the site of protein synthesis. Weissmann and Cantell collected the mRNA from massive quantities of white blood cells. Scientists in Weissmann's lab then copied all of the mRNA back into DNA. By itself, DNA is incapable of replicating, so scientists inserted the DNA into plasmids (DNA that is not part of the chromosomes of an organism, usually a bacterium), which would help the leukocyte DNA to replicate and produce interferon or another protein. The spliced plasmids were customized—or tagged—so that the inserted DNA would be easily retrievable. The scientists then cloned the plasmids in bacteria (*E. coli*) before assaying them for the presence of interferon. This procedure would tell the scientists which DNA carried the gene for interferon. Although the assay the group used often had false-positive results, true-positive results were eventually obtained. More definitive identification of the interferon clones was provided on Christmas Eve Day in 1979, and the patent application was filed soon after.

Given these results, Weissmann's lab continued the interferon work, attempting not only to isolate and describe the structure of the chromosomal interferon gene but also to purify interferon. Today interferon is used to treat a number of diseases, including hairy-cell leukemia, several forms of hepatitis, Kaposi's sarcoma, and malignant melanoma. Some scientists believe, however, that they have just begun to tap the clinical usefulness of interferon.

193

# WILLIAM J. RUTTER (1928– )

Pioneering research in biotechnology has greatly influenced the development of vaccines, most notably for hepatitis B, a virus contracted by over a hundred thousand people each year in the United States. Hepatitis causes inflammation of the liver and in some cases can lead to liver cancer. Today there are seven known hepatitis viruses (A, B, C, D, E, F, and G), some of which are transmitted by contaminated food and water. Hepatitis B, like several of the other hepatitis viruses, is transmitted through bodily fluids. A revolutionary vaccine for hepatitis B was developed by a team of scientists led by William Rutter, in collaboration with both industrial and academic researchers, who used new recombinant DNA techniques.

William Rutter was born in 1928 and grew up in Malad, Idaho. He attended high school there until the age of fifteen, at which point he spent a year at Brigham Young University. Perhaps influenced by stories told by his grandfather, who had been a British army officer in India, Rutter developed an early interest in medicine, particularly in parasitic diseases that abound in warm climates.

After his year at Brigham Young, Rutter claimed to be eighteen and joined the navy, serving until the end of World War II. He then was admitted to Harvard University and received his B.A. in both biochemistry and chemistry in 1949. Once he graduated, in the winter of 1949, Rutter's intention was to enroll in Harvard's medical school the following fall. However, he returned to the West for a few months to be close to his family. During this time he took some medical classes at the University of Utah and did some biochemical research. On the basis of these experiences he quickly determined that his real scientific interest lay in basic research, not in medicine, and so he remained at the University of Utah, working with Gaurth Hansen on studies of metabolism. He earned his M.S. in biochemistry in 1950.

When Hansen was offered a position at the University of Illinois, Rutter traveled with him as a doctoral student. He began to work on galactosemia, a hereditary condition that prevents the normal metabolization of galactose, a constituent of lactose, the sugar found in milk. Infants with this condition are unable to consume milk or milk products, and they risk malnourishment, liver disease, and mental retar-

dation. Rutter received his Ph.D. in biochemistry in 1952 and took up a postdoctoral position at the University of Wisconsin, where he studied enzyme chemistry. His interests led him to take a second postdoctoral position at the Nobel Institute in Stockholm before accepting a teaching position at the University of Illinois.

At Illinois, Rutter developed a greater interest in biological research, and in 1965 he decided to move to Seattle and take a position at the University of Washington, where he learned genetics and began to research mechanisms of gene transcription. Three years later Rutter moved again, this time accepting the chair of the biochemistry department at the University of California at San Francisco (UCSF).

There Rutter became involved in a number of research endeavors, including cloning of the insulin gene. With the advent of recombinant DNA techniques (see Paul Berg et al., p. 179), and the resulting debates within the scientific community about the risks of such experiments, Rutter began to search for a project that would demonstrate the benefits of such technology. He chose production of a hepatitis B vaccine.

Researchers first began to understand the virus during the 1940s, when studies led to a distinction between hepatitis A and B. In the 1960s Baruch Blumberg—later a Nobel laureate—made a discovery that led to a method of identifying hepatitis B carriers. He discovered in the blood of an Australian aborigine an antigen that was rare in North America but prevalent in Africa and Asia, as well as in the blood of individuals who had received a number of transfusions. Over several years Blumberg continued to test blood samples for the presence of this antigen and soon discovered its connection to hepatitis B: it was the surface antigen of the hepatitis B virus. This discovery enabled a specific diagnostic test to be developed.

Rutter and his laboratory made a cooperative arrangement in 1978 with Merck and Company, which at that time also sought a hepatitis B vaccine by using recombinant DNA techniques. Merck had developed a hepatitis B vaccine (Heptavax B) a few years earlier, but it was plasma based, which led to concerns about the potential contamination of plasma—a constituent of blood—by the newly discovered

■ William Rutter, 1991. Courtesy Dr. William Rutter. ©1991 David Powers.

AIDS virus. The vaccine was also extremely difficult and expensive to produce. Because hepatitis cannot be cultivated in a cell culture, the hepatitis B vaccine could not be developed in the same way as the polio vaccine (see Jonas Salk and Albert Bruce Sabin, p. 126) and other subsequent vaccine developments (see Maurice Hilleman, p. 131). However, Roy Vagelos, then president of Merck, Sharp and Dohme Research Laboratories, felt that recombinant DNA technology provided an excellent opportunity, and the project was placed in the area of virus and cell biology research, which Maurice Hilleman headed.

Rutter and Merck agreed to collaborate on a method for producing the hepatitis B antigen in a microbial cell by using recombinant DNA technology. For two years Pablo Valenzuela worked, in Rutter's laboratory, on the necessary gene splicing and by 1981 was able to insert the spliced antigen gene into *E. coli*. When expressed, however, this did not produce antigens that evoked the desired immunologic response. Later it was discovered that the molecules so produced, although equivalent in other respects to the natural antigen, failed because they did not have the proper three-dimensional shape.

In the meantime Rutter was becoming increasingly aware that UCSF could not compete with commercial enterprises, so he suggested to Vagelos that a separate lab be set up to conduct their hepatitis research. Vagelos agreed, and the biotechnology company Chiron Corporation was born. Founded in 1981 by Rutter, Valenzuela, and Edward Penhoet, Chiron engages in a wide variety of activities in such areas as vaccines, therapeutics, and diagnostics and often participates in cooperative ventures with pharmaceutical companies.

The hepatitis researchers learned, soon after the failed *E. coli* attempt, that Genentech and Benjamin D. Hall of the University of Washington had discovered plasmids in baker's yeast. These extrachromosomal DNA molecules capable of replicating independently had previously been known to exist only in bacteria. Researchers felt that this discovery would provide a possible method of producing the antigen. Further, culturing yeast in large quantities was a well-known technology that would lend itself relatively easily to scaling up production. The project then expanded to include Merck, Chiron, Rutter and his UCSF laboratory, and Hall and his laboratory. These collaborators were able to produce plasmids that included the DNA sequence of the hepatitis B antigen. The plasmids were transferred to yeast, which then produced the desired antigen. The antigen in turn stimulated the correct antibody reaction. The researchers had solved their problem.

Clinical studies were done on the new hepatitis B vaccine, first on volunteers among the Merck employees and then on a larger scale. These studies concluded that the new vaccine provided protection comparable to that of Heptavax-B, and in 1986 Recombivax HB was licensed in the United States. Today the vaccine is administered to infants, children under the age of eighteen, and adults at risk for contracting the virus.

# PART IV.

# FINDING
# TOMORROW'S
# MEDICINES

The medicines now in the pipeline to become tomorrow's medicines and those to be developed in the future will inevitably depend in part on the advances made by the scientists whose careers and contributions have been discussed in the preceding chapters. It is still too early to identify the "achievers" in the present whose contributions will in retrospect be considered key or exemplary, but their achievements will probably fall in one of several areas of research that have opened up in very recent times—some with clear links to the past. Meanwhile, researchers are already using new methods of finding pharmaceuticals—principally computational and combinatorial chemistry and high-throughput screening—that pharmaceutical achievers of the future will almost certainly employ.

# 15. New or Revisited Areas for Pharmaceutical Research

## Gene Therapy

The completion of the project to understand the human genome in terms of the genetic codes contained on each of its twenty-three pairs of chromosomes opens the door to the development of many more biotechnological therapies. In the future more diseases and susceptibilities to disease will be linked to the malfunctioning of specific genes—or more likely, to combinations of genes and environmental factors. The time it takes scientists to locate such disease-related genes will be greatly reduced by ready reference to computer databases compiled during the project. Scientists will even be able to extrapolate from the genome to the disease. Before the human genome was mapped, they had to begin with knowledge of the protein and then work back to the genome, as in developing the biotech versions of insulin, erythropoietin, and interferon. Scientists inspecting the genome have already recognized a code for a third type of histamine receptor whose role they do not yet understand, but they remember well the discovery of a second type of histamine receptor, which resulted in a revolutionary treatment for ulcers, cimetidine (Tagamet) (see James Whyte Black, p. 148).

As a result of detailed knowledge of the human genome, the now-familiar sort of biotechnology—involving the insertion of human genes into viruses or other microorganisms that are in turn coaxed in the laboratory to express these genes as critical proteins, such as human insulin—will become even more common.

It will also be possible to introduce normal genes in vitro into human tissue cells that would ordinarily express a particular gene, except that the gene is either absent or mutated in an individual, and then reinsert the modified tissue cells into that individual. The new, fully functioning gene is then expressed instead of the faulty gene, thus treating the individual's condition. The genetic "mistake" is not completely corrected, however, and could still be passed on to any offspring. Furthermore, this method requires repeated dosing with the normal gene. It is being tested for use in the treatment of a number of diseases, including cystic fibrosis, sickle-cell anemia, and Duchenne's muscular dystrophy. A similar method is also being tested for treatment of late-stage tumors.

Another form of gene therapy is being used to treat other serious medical conditions. Here the gene used bears no causal relationship to the condition but works to repair the effects of the disease. For example, in the case of heart disease, a gene that stimulates the growth of blood vessels has been injected into heart muscle. Once new blood vessels begin to grow, blood can flow to the heart without passing through clogged arteries.

A more controversial mode of gene therapy involves introducing a normal gene into an early embryonic cell that possesses a faulty gene. This method produces a full cure, correcting the faulty gene, and therefore ensures that offspring will also be free of the genetic "mistake." This type of gene therapy raises serious ethical questions, however, as the manipulations affect an entire genetic line, not just a single individual. There is understandable concern about what might be considered a "faulty" gene in years to come and who would make such decisions.

## Antibiotics

Another area of great research activity today centers on antibiotics, but unlike gene therapy, this research represents a return to an earlier direction. After the development of penicillin and other now-standard antibiotics used to treat infections and many infectious diseases of all kinds, pharmaceutical companies tended to focus instead on chronic disease research. With the recent realization that increasing numbers of bacteria are developing resistance to standard antibiotics, pharmaceutical companies are reentering the field, creating new and better antibiotics to combat the ever-strengthening bacteria.

Bacteria can resist drugs in three ways: they can prevent the drug from reaching its target; they can destroy or modify the drug so that it cannot bind to its target; or they can alter their own structure so that they no longer serve as the target for a particular antibiotic. Bacteria become resistant to antibiotics both through genetic mutations and by acquiring resistance genes from other bacteria. Bacteria can trade genetic material by exchanging plasmids—genetic material in the cell that is not part of its chromosomes.

Antibiotic resistance has become a greater problem in recent years because of the imprudent use of this class of drugs. For example, antibiotics are frequently prescribed to treat viral infections that will not respond to such drugs. Further, patients often fail to take an entire cycle of antibiotics. These practices lead to increased antibiotic resistance because they facilitate the demise of the weakest bacteria, leaving the stronger and more resistant bacteria to survive with minimal competition. And some bacteria are becoming resistant to more than one antibiotic, which adds to the pressure on pharmaceutical companies to produce different and stronger antibiotics. To a significant extent the companies are succeeding, creating in some cases whole new classes of antibiotics.

## Orphan Drugs

Pharmaceutical companies, as commercial endeavors, typically develop drugs they feel will be profitable. Unfortunately these economic considerations left many rare diseases without treatment until the Orphan Drug Act of 1983 provided financial support and other incentives for pharmaceutical companies to move into such areas. In the early 1980s a woman who was unable to obtain drugs—not even by having a friend carry them across the border from Canada—to alleviate her son's Tourette's syndrome (characterized by grimacing and blinking and often loud, uncontrollable vocalizations) made an appeal to her congressman,

California's Henry Waxman. He was moved by the appeal and held hearings that ultimately resulted in the Orphan Drug Act. The particular circumstances of the incident—pimozide being carried across the Canadian border—also contributed to legalizing the act of bringing non–FDA-approved medicines into the United States, provided they are for personal use.

According to the Orphan Drug Act, an orphan drug must meet at least one of the following two criteria: it is intended to treat a disease affecting fewer than two hundred thousand people in the United States; and it will not be profitable within seven years after its approval by the FDA. Orphan diseases are typically chronic and therefore require lifelong

■ A child examines a structural model of DNA. © Corbis.

treatment; they are also often life threatening. An example of an orphan drug is erythropoietin (see George Rathmann and Fu-Kuen Lin, p. 187), which treats the anemia of end-stage kidney disease, a condition that affects an estimated 192,000 people in the United States. Diseases like leprosy, tuberculosis, and malaria would qualify for orphan drugs because they are relatively rare in the United States, although they continue to afflict many hundreds of thousands of people worldwide.

The Orphan Drug Act provides a number of incentives for pharmaceutical companies to develop these important treatments. Financial incentives have included grants to subsidize clinical trials or tax credits based on the cost of the trials to a company and waivers of some of the fees associated with filing for FDA approval. FDA staff also help speed the process of application, putting drugs on the fast track for approval because many orphan diseases are life threatening. Finally, the act provides for seven years of exclusive marketing by the drug's developer, which is an important provision because many of the substances, such as erythropoietin and the AIDS drug azidothymidine (AZT), have been in the public domain so long that they cannot be patented as products.

## IMPROVED CHEMICAL ENTITIES

Considerable pharmaceutical research focuses on improving existing drugs to make them more effective and safer. Indeed, some companies' entire business revolves around making improvements to existing drugs. One way of improving a drug—actually a method that has been used for a long time—is to look at its metabolites, the compounds into which it is converted in the human body. One of these compounds may be the truly effective agent, which may have significantly better properties than the compound that was initially ingested or injected.

A more recent concern has been to look for safer or the most effective stereoisomers of compounds that exhibit the capacity, quite common in complex organic compounds, to rotate plane-polarized light to the left or to the right because of the particular spatial arrangement of identical atomic groupings in the compound (see Louis Pasteur, p. 6, for a discussion of optical activity). An early example of the

recognition that only one of a pair of isomers constitutes effective treatment of a disease occurred in the development of L-dopa for Parkinson's disease, which is characterized by the shaking of resting muscles. The earliest clinical trials were conducted in the 1960s, using a mixture of left- and right-rotating dopa, the precursor of dopamine in the brain. This mixture could be extracted from broad beans (*Vicia faba*) or synthesized in the laboratory. Then it was discovered that the brain could only metabolize L-dopa, and it made sense to administer just this isomer, since the presence of the other clearly weakened the dose being delivered and might prove harmful. In general, stereoisomers can be separated by physical or chemical means, or in some cases syntheses can be designed so that only the preferred isomer is produced—processes that add to the cost of manufacturing various drugs.

## DRUG DELIVERY

Designing new ways to deliver drugs to the sites where they are needed in the body has become more important than ever before with the advent of biotechnology, its capability of reproducing the many large molecules that the body normally uses to regulate itself, and the therapeutic use of such molecules. These large molecules are often barred from entry into various of the body's systems or are quickly and easily destroyed by the body's own defenses—even when injected into the bloodstream. Older therapies involving small molecules have also been foiled in some instances by such mechanisms as the blood-brain barrier.

Scientists have developed synthetic polymer nanospheres as one means of surmounting these problems. The nanospheres are microscopic, spherical containers made of a polymer core covered by a protective coating made of a different polymer that can, if desired, be tailored to fit into special receptors on certain cells. Drugs are dispersed in the polymer core, and the nanospheres can be injected intravenously. Because of the protective coating that dissolves slowly, they can evade destruction and are capable of remaining in the bloodstream for hours. The coating can also help target the drugs to specific organs. Such nanospheres are already being used, for example, to deliver chemotherapy directly to brain tumors.

Deep-lung drug delivery is another method that promises to transport large molecules into the bloodstream without destroying them. With this method patients do not have to endure the multiple injections that control of chronic disease usually requires. The aerosols that have long been on the market to treat asthma deliver drugs locally, not to the deep lung, and per puff they do not deliver much drug mass. In deep-lung delivery the inhaling device converts powdered drugs into aerosols, rather than using liquid aerosols, thus making more concentrated doses possible. Powdered drugs not only are more effective; they also have relatively long shelf lives and resist the growth of bacteria and other damaging microbes. Medications for diabetes, osteoporosis, emphysema, multiple sclerosis, and infertility are all being investigated as potential candidates for inhalants.

The future promises to bring many more unusual systems for effecting complex controlled release of many kinds of medicines in the body. One already in the works involves implanting or having the patient swallow a silicon microchip containing tiny reservoirs of medicine covered with thin films of gold. The medicine is released when the gold, serving as the anode of a tiny electrolytic cell, is dissolved in the body's saline-solution environment. The necessary electric potential between anode and cathode would presumably be set up by a signaling device inside or outside the body.

■ Bruce Merrifield with a peptide synthesizer.
Courtesy Bruce Merrifield.

# 16. New Methods of Drug Discovery

## COMBINATORIAL CHEMISTRY

In recent times the methods of looking for cures have advanced beyond those discussed in previous chapters. The hope is that these new methods will reduce both the manpower and the time required to find new medicines and thus reduce their price to consumers.

Combinatorial chemistry has brought mass production to the search for lead candidates in the drug discovery process. When scientists feel the need to synthesize new compounds, however, they are freed from the time-consuming and expensive process of synthesizing one compound at a time—an approach used at least since Paul Ehrlich's then-astonishing synthesis in the first decade of the twentieth century of three hundred compounds in three years before finding salvarsan (see Ehrlich, p. 17). Now thousands of compounds that are systematic variations of a particular molecular structure can be generated quickly.

The technology of combinatorial chemistry, which has applications in many chemical fields, traces back to the automatic synthesizing machines that Bruce Merrifield, professor of biochemistry at Rockefeller University, built in the 1960s. Key to the success of his method of having a machine perform the tedious steps of adding each amino acid to a growing peptide chain was the idea of using tiny porous beads of polystyrene to which he attached the first amino acid. The bead kept the growing chain from washing away during purification steps. After adding the last desired amino acid, the new polypeptide chain could be removed from the bead by using the proper reagents. Known as "solid-phase" synthesis, this method was extended to building other kinds of molecules, and computer control was introduced to replace the comparatively primitive electrical switching devices originally used to perform such functions as moving along the reaction vessel, releasing reagents, and controlling temperature. Today many automatic synthesizers operate solely in the liquid phase, which is more versatile but presents more difficulties in separating products from reagents and by-products.

■ A Gene Chip probe array consisting of gene sequences—as many as twelve thousand—in precise positions on a glass chip. Among their many uses, DNA chips can be set up to test the toxicity on the genome of potential drugs. Courtesy Affymetrix, Inc.

Whereas Merrifield's original objective was simply to produce one product at a time more efficiently, multiplication of the vessels containing reagents and their robotic manipulation enables the generation of hundreds and even

■ Leonard Skeggs with his automatic method for colorimetric analysis. Courtesy Veterans Affairs Medical Center, Brecksville, Ohio.

thousands of compounds that are all combinations of a selected group of molecular structures. In one method, called "split" or "pool-and-split," after each stage of the synthesis the products are mixed together and then split into batches for further synthesis. Screening must then take place to find the batches and ultimately the individual compounds that display some slight reactivity with the chosen biological target. Another method, known as "parallel synthesis," creates chemical structure combinations separately but in parallel. For example, all members of a set of reagents A are added to a set of reagents B, which are set out in small wells in a plate of ninety-six or more wells, in order to make all possible products of AB.

In the early stages of a search for a drug candidate, especially in cases in which little is known of the structure of the receptor, researchers may try to synthesize a great diversity of molecular structures, looking for a few that show some of the desired reactivity. After such compounds are located, they then synthesize whole libraries of compounds that are structurally very similar to each other.

## COMPUTATIONAL CHEMISTRY

Given the endless possible molecular structures, a great deal of planning must go into setting up the syntheses. To support this planning, researchers usually turn to computational chemistry—chemistry based on computers. Databases and computer programs make it possible to design a chemical substance that has particular properties, for example, one that will bind to what is presumed to be an active site on a target molecule. Discovering the active site itself may also have depended on computer methods. In some instances little is known about the three-dimensional structure of the target, but from just the sequence of the atoms (a two-dimensional description) the computer can produce likely three-dimensional models by using quantum mechanical calculation or by comparison with the structures of other similarly sequenced molecules.

## NEW METHODS OF SCREENING

Even before turning to actual syntheses, researchers can design and conduct "virtual" screening of potential drug leads on the computer. This process is sometimes called "in silico" screening in contrast to the classical methods of in vitro (in laboratory glassware) and in vivo (in animals or humans).

Just as combinatorial and computational chemistry was being developed, high-throughput screening (HTS) appeared on the scene and became an integral component of today's search for new pharmaceuticals. HTS involves the use of highly automated, very fast techniques of in vitro testing of compounds that trace back to the autoanalyzer developed in the 1950s by Leonard Skeggs, a biochemist employed by the Veterans Administration hospital in Parma, Ohio. The clinical laboratory he supervised was having difficulty keeping up with its workload of blood tests ordered by physicians. At that time technicians approached

any multistep operation as a batch process: once the final step of a test had been performed on a group of samples, they turned to the next batch, beginning again with the first step. To replace this process, Skeggs designed an automated continuous process for two common tests—those of blood urea nitrogen levels and blood sugar levels. Technicians needed only load the rotating disks with test tubes of blood, which could be inserted at any time, and later gather the printouts from the analytical instruments—in this case colorimeters, which compare the degree of color change in each sample with the color of reference samples. Between these two steps the automatic techniques took small amounts of blood from the original samples and removed from these small portions the compounds that were not relevant to the test in question. The appropriate reagents were then added to the remainder to produce color changes, and the products of these reactions were run through the automatically recording colorimeters.

Today's autoanalytical devices are far more elaborate and faster than Skeggs's prototypes and can handle hundreds of thousands of analyses per day. In HTS similar devices are used to determine the molecular structures of compounds found in molecular libraries—either those collected over time or those just created by the processes of combinatorial chemistry. Among the analytical techniques used are infrared spectroscopy, nuclear magnetic resonance, mass spectroscopy, and liquid chromatography. Screening procedures also encompass bioassays to see whether and how various compounds will be metabolized in the human body, whether they will bind to the chosen target, and whether they will prove toxic to human beings.

One method of determining toxicity involves the use of DNA chips. DNA chips are arrays of DNA fragments, mounted on small glass wafers, that can be engineered to eliminate the primary target of a particular drug candidate. Such chips are exposed to the drug, and the effects are analyzed. Because the primary target is missing from the chip, secondary or side effects of the drug can be observed—harmful ones, which need to be guarded against, and potentially useful ones, which indicate that the drug could be used for other therapeutic purposes.

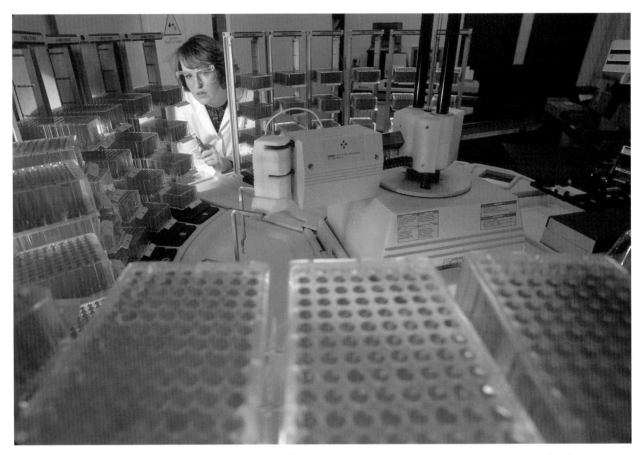

■ A Pfizer scientist in the company laboratory in Sandwich, England, observes high-throughput screening of large numbers of compounds for potential drug candidates. Courtesy Pfizer, Inc.

## MEDICINES IN THE DAYS AFTER TOMORROW?

As we enter the third millennium, new opportunities in pharmaceutical research and development abound, both in areas to be investigated and in methods of discovery. But if history teaches anything, it is that opportunities will arise in now barely imagined areas—just as antibiotics or biotechnology arrived without much precedent (although each of these fields certainly had some antecedents to which historians can point). Further innovations are likewise to be expected in the methods of finding new pharmaceuticals, and these and other innovations will be accomplished by scientists who have just begun their careers or are only now considering the exciting possibilities of a life in pharmaceuticals.

# Bibliography

For writings on scientists besides those listed under the individual chapter titles, please refer to General Histories and Overviews, Collected Biographies, and Oral Histories at the end of this bibliography.

## 1. EARLY PHARMACEUTICALS

**Ernst Bäumler.** *Paul Ehrlich: Scientist for Life.* Translated by Grant Edwards. New York: Holmes and Meder Publishers, 1984.

**Günter Benz; Ralf Hahn; Carsten Reinhardt.** *100 Jahre Chemisch-Wissenschaftliches Laboratorium der Bayer AG in Wuppertal-Elberfeld 1896–1996.* Leverkusen, Germany: Bayer AG, 1996.

**Joseph C. Collins; John R. Gwilt.** "The Life Cycle of Sterling Drug, Inc." *Bulletin for the History of Chemistry* 25 (2000), 22–27.

**Patrice Debré.** *Louis Pasteur.* Translated by Elborg Forster. Baltimore: Johns Hopkins University Press, 1998.

**Marcel Delépine.** "Joseph Pelletier and Joseph Caventou." *Journal of Chemical Education* 28 (Sept. 1951), 454–461.

**René Dubos.** *Pasteur and Modern Science.* Garden City, N.Y.: Anchor Books, 1960.

**Gerald Geison.** *The Private Science of Louis Pasteur.* Princeton, N.J.: Princeton University Press, 1995.

**Harry Henderson.** "The Paternity of Aspirin." *Science* 286 (5 Nov. 1999), 1090.

**Constance Holden.** "Aspirin's Paternity Debated." *Science* 186 (1 Oct. 1999), 39.

**Charles C. Mann; Mark L. Plummer.** *The Aspirin Wars: Money, Medicine, and 100 Years of Rampant Competition.* New York: Alfred A. Knopf, 1991.

**Jan R. McTavish.** "Aspirin in Germany: The Pharmaceutical Industry and the Pharmaceutical Profession." *Pharmacy in History* 29 (1987), 103.

**J. M. D. Olmstead.** *François Magendie: Pioneer in Experimental Physiology and Scientific Medicine in XIX Century France.* New York: Schuman's, 1944.

**Arthur M. Silverstein.** *Paul Ehrlich's Receptor Immunology: The Magnificent Obsession.* San Diego: Academic Press, 2001.

## 2. EARLY PHARMACEUTICAL COMPANIES AND THE PURE FOOD AND DRUGS ACT

**Oscar E. Anderson, Jr.** *The Health of a Nation: Harvey W. Wiley and the Fight for Pure Food.* Chicago: University of Chicago Press, 1958.

**Lawrence G. Blochman.** *Doctor Squibb: The Life and Times of a Rugged Idealist.* New York: Simon and Schuster, 1958.

**John Francis Marion.** *The Fine Old House: SmithKline Corporation's First 150 Years.* Philadelphia: SmithKline Corporation, 1980.

**Harvey W. Wiley.** *An Autobiography.* Indianapolis: Bobbs-Merrill Company, 1930.

**James Harvey Young.** *Pure Food: Securing the Federal Food and Drugs Act of 1906.* Princeton, N.J.: Princeton University Press, 1989.

## 3. VITAMINS

**Rima Apple.** *Vitamania: Vitamins in American Culture.* New Brunswick, N.J.: Rutgers University Press, 1996.

**Jeffrey Sturchio,** ed. *Values and Visions: A Merck Century.* Rahway, N.J.: Merck & Company, 1991.

**Louis Haber.** "Gladys Anderson Emerson: Biochemist Whose Researches with Experimental Animals Have Increased Our Knowledge of the Effects of Vitamin Deficiencies on the Human Body." In *Women Pioneers of Science*, pp. 140–155. New York: Harcourt Brace Jovanovich, 1979.

**István Kardos,** interviewer. "A Talk with Albert Szent-Györgyi." *New Hungarian Quarterly* 16 (1975), 136–150.

**Ralph W. Moss.** *Free Radical: Albert Szent-Györgyi and the Battle over Vitamin C.* New York: Paragon House, 1988.

**Richard L. Mueller; Stephen Scheidt.** "History of Drugs for Thrombotic Disease: Discovery, Development, and Directions for the Future." *Circulation* 89:1 (1994), 432–449.

*A National Historic Chemical Landmark: Albert Szent-Györgyi and the Discovery of Vitamin C.* Szeged, Hungary, 11 May 2002. Washington, D.C.: American Chemical Society Office of Communications, 2002. See on-line version at www.chemistry.org/landmarks.

**Robert R. Williams.** "The Beriberi Vitamin." *Industrial and Engineering Chemistry* 29 (Sept. 1937), 980–984.

**Edna Yost.** "Dr. Gladys Anderson Emerson: Biochemist and Nutritionist." In *Women of Modern Science*, pp. 73–82. New York: Dodd, Mead, 1959.

## 4. ENZYMES, HORMONES, AND NEUROTRANSMITTERS

**Michael Bliss.** *The Discovery of Insulin.* Chicago: University of Chicago Press, 1982.

———. "The Discovery of Insulin: The Inside Story." In *The Inside Story of Medicines: A Symposium,* edited by Gregory J.

Higby and Elaine C. Stroud, pp. 93–99. Madison, Wis.: American Institute of the History of Pharmacy, 1997.

**W. S. Feldberg.** "The Early History of Synaptic and Neuromuscular Transmission by Acetylcholine: Reminiscences of an Eye Witness." In *The Pursuit of Nature: Informal Essays on the History of Physiology,* by A. L. Hodgkin et al., pp. 65–83. New York: Cambridge University Press, 1977.

**Joseph Fruton.** *Contrasts in Scientific Style: Research Groups in the Chemical and Biochemical Sciences.* Philadelphia: American Philosophical Society, 1990.

**Seale Harris.** *Banting's Miracle: The Story of the Discoverer of Insulin.* Philadelphia: Lippincott, 1946.

**E. J. Kahn, Jr.** *All in a Century: The First 100 Years of Eli Lilly and Company.* West Cornwall, Conn.: Kahn, 1975.

**Jane H. Murnaghan; Paul Talalay.** "H. H. Dale's Account of the Standardization of Insulin." *Bulletin of the History of Medicine* 66 (1992), 440–450.

**Abigail O'Sullivan.** "Henry Dale's Nobel Prize Winning 'Discovery.' " *Minerva* 39 (2001), 409–424.

**John Parascandola.** *The Development of American Pharmacology: John J. Abel and the Shaping of a Discipline.* Baltimore: Johns Hopkins University Press, 1992.

*Takamine: Documents from the Dawn of Industrial Biotechnology.* Elkhart, Ind.: Miles, Inc., 1988.

## 5. Sulfa Drugs

**Daniel Bovet.** "The Relationships between Isosterism and Competitive Phenomena in the Field of Drug Therapy of the Autonomic Nervous System and That of the Neuromuscular Transmission." In *Nobel Lectures in Physiology or Medicine,* vol. 3, pp. 552–578. Amsterdam: Elsevier, 1964.

**John Lesch.** "Chemistry and Biomedicine in an Industrial Setting: The Invention of the Sulfa Drugs." In *Chemical Sciences in the Modern World,* edited by Seymour H. Mauskopf, pp. 158–215. Philadelphia: University of Pennsylvania Press, 1993.

## 6. Antibiotics

**Richard S. Baldwin.** *The Fungus Fighters: Two Women and Their Discovery.* Ithaca, N.Y.: Cornell University Press, 1981.

**Lennard Bickel.** *Howard Florey: The Man Who Made Penicillin.* Victoria, Australia: Melbourne University Press, 1972.

**Ronald W. Clark.** *The Life of Ernst Chain: Penicillin and Beyond.* New York: St. Martin's, 1985.

**Albert L. Elder.** *The History of Penicillin Production.* Chemical Engineering Progress Symposium Series, no. 100. New York: American Institute of Chemical Engineers, 1970.

**Georgina Ferry.** *Dorothy Hodgkin: A Life.* London: Granta Books, 1998.

**Gladys L. Hobby.** *Penicillin: Meeting the Challenge.* New Haven, Conn.: Yale University Press, 1985.

*An International Historic Chemical Landmark: The Discovery and Development of Penicillin, 1928–1945.* The Alexander Fleming Laboratory Museum, London, 19 November 1999. Washington, D.C.: American Chemical Society Office of Communications; London: Royal Society of Chemistry, 1999.

**Sharon Bertsch McGrayne.** "Dorothy Crowfoot Hodgkin." In *Nobel Prize Women in Science: Their Lives, Struggles, and Momentous Discoveries.* New York: Birch Lane Press, 1993, 225–254.

**Carol L. Moberg; Zanvil A. Cohn,** eds. *Launching the Antibiotic Era: Personal Accounts of the Discovery and Use of the First Antibiotics.* New York: Rockefeller University Press, 1990.

**John Parascandola,** ed. *The History of Antibiotics: A Symposium.* Madison, Wis.: American Institute of the History of Pharmacy, 1980.

**Milton Wainwright.** *Miracle Cure: The Story of Penicillin and the Golden Age of Antibiotics.* Cambridge, Mass.: Basil Blackwell, 1990.

**Selman A. Waksman.** *My Life with the Microbes.* New York: Simon and Schuster, 1954.

## 7. Steroids

**Carl Djerassi.** *The Pill, Pygmy Chimps, and Degas' Horse.* New York: Basic Books, 1992.

————. *Steroids Made It Possible.* Profiles, Pathways, and Dreams, series editor Jeffrey I. Seeman. Washington, D.C.: American Chemical Society, 1990.

*An International Historic Chemical Landmark: The "Marker Degradation" and Creation of the Mexican Steroid Hormone Industry, 1938–1945.* University Park, Pennsylvania, 1 October 1999; Mexico City, 2 December 1999. Washington, D.C.: American Chemical Society Office of Communications; Mexico City: Sociedad Quimica de Mexico, 1999.

**G. Hetenyi, Jr.; J. Karsh.** "Cortisone Therapy: A Challenge to Academic Medicine in 1949–1952." *Perspectives in Biology and Medicine* 40 (1997), 426–439.

**Edward C. Kendall.** *Cortisone.* New York: Charles Scribner's Sons, 1971.

Pedro A. Lehmann; Antonio Bolivar; Rodolfo Quintero. "Russell E. Marker." *Journal of Chemical Education* 50 (1973), 195–199.

Russell Marker. "The Early Production of Steroidal Hormones." *CHOC News* 4:2 (Summer 1987), 3–6.

Wendy B. Murphy. *Science and Serendipity: A Half Century of Innovation at Syntex.* White Plains, N.Y.: Benjamin Company, 1994.

*A National Historic Chemical Landmark: Synthesis of Physostigmine.* DePauw University, Greencastle, Indiana, 23 April 1999. Washington, D.C.: American Chemical Society Office of Communications, 1999. See on-line version at www.chemistry.org/landmarks.

## 8. DRUGS AFFECTING THE CENTRAL NERVOUS SYSTEM: ANTIPSYCHOTICS AND TRANQUILIZERS

Otto Theodor Benfey; Peter J. T. Morris, eds. *Robert Burns Woodward: Architect and Artist in the World of Molecules.* Philadelphia: Chemical Heritage Foundation, 2001.

Mary Ellen Bowden; Theodor Benfey. *Robert Burns Woodward and the Art of Organic Synthesis.* Philadelphia: Beckman Center for the History of Chemistry, 1992.

Anne E. Caldwell. *Origins of Psychopharmacology from CPZ to LSD.* Springfield, Ill.: Charles C. Thomas, 1970.

John Francis Marion. *The Fine Old House: SmithKline Corporation's First 150 Years.* Philadelphia: SmithKline Corporation, 1980.

J.-P. Olie; D. Ginestet; G. Jolles; H. L. Lôo, eds. *Historie d'une découverte en psychiatrie: 40 ans de chimiothérapie neuroleptique.* Paris: Fondation Rhône-Poulenc Rorer, Centre Hospitalier-Sainte-Anne, 1992.

Daniel M. Perrine. *The Chemistry of Mind-Altering Drugs: History, Pharmacology, and Cultural Context.* Washington, D.C.: American Chemical Society, 1996.

Public Histoire (Fabienne Gambrelle). *Innovating for Life: Rhône-Poulenc, 1895–1995.* Translated by Elizabeth A. Kahn and Joseph S. Kahn. Paris: Public Histoire Albin Michel, 1995.

Leo B. Slater. "Organic Synthesis and R. B. Woodward: An Historical Study in the Chemical Sciences." Ph.D. diss., Princeton University, 1997.

Mickey C. Smith. *A Social History of the Minor Tranquilizers: The Quest for Small Comfort in the Age of Anxiety.* New York: Pharmaceutical Products Press, 1991.

L. H. Sternbach. *The Benzodiazepine Story.* Basel, Switzerland: Hoffmann–La Roche, 1983.

Judith P. Swazey. *Chlorpromazine in Psychiatry: A Study of Therapeutic Innovation.* Boston: MIT Press, 1974.

Jean Thuillier. *Ten Years That Changed the Face of Mental Illness.* Translated by Gordon Hickish. London: Martin Dunitz, 1999.

Glenn E. Ullyot; Maxwell Gordon. "Psychopharmacological Agents." In *Kirk-Othmer Encyclopedia of Chemical Technology*, second edition, vol. 16, pp. 640–679. New York: Interscience, 1968.

## 9. NEWER VACCINES

Huntly Collins. "The Man Who Changed Your Life: Maurice Hilleman's Vaccines Prevent Millions of Deaths Every Year." *Inquirer Magazine*, 29 August 1999, pp. 6–12.

Louis Galambos, with Jane Sewell. *Networks of Innovation: Vaccine Development at Merck, Sharp & Dohme, and Mulford, 1895–1995.* New York: Cambridge University Press, 1995.

Naomi Rogers. *Dirt and Disease: Polio before FDR.* New Brunswick, N.J.: Rutgers University Press, 1992.

Jane S. Smith. *Patenting the Sun: Polio and the Salk Vaccine.* New York: William Morrow and Company, 1990.

Jeffrey Sturchio, ed. *Values and Visions: A Merck Century.* Rahway, N.J.: Merck and Company, 1991.

## 10. CLINICAL TESTING AND REFORM OF THE 1906 PURE FOOD AND DRUGS ACT

Charles O. Jackson. *Food and Drug Legislation in the New Deal.* Princeton, N.J.: Princeton University Press, 1970.

Harry M. Marks. *The Progress of Experiment: Science and Therapeutic Reform in the United States, 1900–1990.* New York: Cambridge University Press, 1997.

Richard E. McFadyen. "Thalidomide in America: A Brush with Tragedy." *Clio Medica* 11 (1976), 79–93.

Mark Parascandola. "Clinical Trials: New Developments and Old Problems." In *The Inside Story of Medicines: A Symposium*, edited by Gregory J. Higby and Elaine C. Stroud, pp. 201–214. Madison, Wis.: American Institute of the History of Pharmacy, 1997.

*Statistics in Medicine* 1:4 (Oct.–Dec. 1982). The issue was published as an appreciation of Sir Austin Bradford Hill.

John Swann. "Sure Cure: Public Policy on Drug Efficacy before 1962." In *The Inside Story of Medicines: A Symposium*, edited by Gregory J. Higby and Elaine C. Stroud, pp. 223–261. Madison, Wis.: American Institute of the History of Pharmacy, 1997.

Suzanne White. "The Chemogastric Revolution and the Regulation of Food Chemicals." In *Chemical Sciences in the Modern World*, edited by Seymour Mauskopf, pp. 322–355. Philadelphia: University of Pennsylvania Press, 1993.

James Harvey Young. *The Medical Messiahs: A Social History of Health Quackery in Twentieth-Century America.* Princeton, N.J.: Princeton University Press, 1992.

———. "Sulfanilamide and Diethylene Glycol." In *Chemistry and Modern Society*, edited by John Parascandola and James C. Whorton, pp. 105–125. Washington, D.C.: American Chemical Society, 1983.

## 11. Recent Cardiovascular and Antiulcer Drugs

John Diebold. "Mevacor: Developing the Weapon against Cholesterol." In *The Innovators: The Discoveries, Inventions, and Breakthroughs of Our Time*, pp. 152–176. New York: E. P. Dutton, 1990.

John V. Duncia et al. "The Discovery of DuP 753, a Potent, Orally Active Nonpeptide Angiotensin II Receptor Antagonist." *Medicinal Research Reviews* 12 (Mar. 1992), 149–191.

C. Robin Ganellin. "Cimetidine." In *Chronicles of Drug Discovery*, edited by Jasjit S. Bindra and Daniel Lednicer, vol. 1, pp. 1–38. New York: John Wiley, 1982.

*An International Historic Chemical Landmark: The Discovery of Histamine $H_2$-Receptor Antagonists.* SmithKline Beecham Pharmaceuticals, Harlow, U.K., and King of Prussia, Pennsylvania, 24 November 1997 and 27 February 1998. Washington, D. C.: American Chemical Society; London: Royal Society of Chemistry, 1997.

Alexander L. Johnson et al. "Nonpeptide Angiotensin II Receptor Antagonists." *Drug News & Perspectives* 3:6 (July 1990), 337–351.

John Francis Marion. *The Fine Old House: SmithKline Corporation's First 150 Years.* Philadelphia: SmithKline Corporation, 1980.

Miguel Ondetti. "From Peptides to Peptidases: A Chronicle of Drug Discovery." *Annual Review of Pharmacology* 34 (1994), 1–16.

Miguel A. Ondetti; David W. Cushman; Bernard Rubin. "Captopril." In *Chronicles of Drug Discovery*, edited by Jasjit S. Bindra and Daniel Lednicer, vol. 2, pp. 1–31. New York: John Wiley, 1983.

Irwin Ross. "A New Drug That Fights Cholesterol." *Reader's Digest* 133:788 (Dec. 1987), 91–94.

Dean Stanley Tarbell; Ann Tracy Tarbell. *Essays on the History of Organic Chemistry in the United States, 1875–1955.* Nashville, Tenn.: Folio Publishers, 1986, 373–382.

P. Roy Vagelos. *Promise and Performance in Pharmaceutical Research: Merck as a Case Study in Continuous Technological Evolution.* Philadelphia: Chemical Heritage Foundation, 1999. An address delivered at the Tenth Annual Othmer Luncheon, at which Dr. Vagelos was awarded the third Othmer Gold Medal.

Rein Vos. *Drugs Looking for Diseases: Innovative Drug Research and the Development of the Beta Blockers and the Calcium Antagonists.* Boston: Kluwer Academic Publishers, 1991.

## 12. Recent Antihistamines and Antidepressants

Peter Breggin; Ginger Ross Breggin. *Talking Back to Prozac: What Doctors Won't Tell You about Today's Most Controversial Drug.* New York: St. Martin's Press, 1994.

David A. Karp. *Speaking of Sadness: Depression, Disconnection, and the Meanings of Illness.* New York: Oxford University Press, 1996.

Peter Kramer. *Listening to Prozac.* New York: Viking, 1993.

Daniel M. Perrine. *The Chemistry of Mind-Altering Drugs: History, Pharmacology and Cultural Context.* Washington, D.C.: American Chemical Society, 1996.

Edward Shorter. *A History of Psychiatry: From the Era of the Asylum to the Age of Prozac.* New York: John Wiley & Sons, 1997.

## 13. Cancer and AIDS Drugs

Susan Ambrose et al. "Gertrude Belle Elion." In *Journeys of Women in Science and Engineering: No Universal Constant.* Philadelphia: Temple University Press, 1997, pp. 135–140.

Louis Galambos. "War against HIV." *Invention & Technology* 15 (Spring 2000), 56–63.

Louis Galambos; Jane Eliot Sewell. *Confronting AIDS: Science and Business Cross a Unique Frontier.* Whitehouse Station, N.J.: Merck & Company, 1999.

Jiunn H. Lin; Drazin Ostovic; Joseph P. Vacca. "The Integration of Medicinal Chemistry, Drug Metabolism, and Pharmaceutical Research and Development in Drug Discovery and Development: The Story of Crixivan, an HIV Protease Inhibitor." In *Integration of Pharmaceutical Discovery and Development*, edited by Ronald T. Borchardt et al., pp. 233–255. New York: Plenum Press, 1998.

Sharon Bertsch McGrayne. "Gertrude Elion." In *Nobel Prize Women in Science: Their Lives, Struggles, and Momentous Discoveries*, pp. 280–303. New York: Birch Lane Press, 1993.

14. BIOTECHNOLOGY

**Lisa Bain.** "First Clone" [about Stanley Cohen]. *Penn Medicine* 28 (Winter 1990/1), 29–33.

**James Bates.** "Biotech Detective Scores Coup" [about Fu-Kuen Lin]. *Los Angeles Times*, 2 June 1989, Part I, pp. 1, 5.

**Alberto Cambrosio; Peter Keating.** *Exquisite Specificity: The Monoclonal Antibody Revolution.* New York: Oxford University Press, 1995.

*The Emergence of Biotechnology: DNA to Genentech* [transcript and videotape of symposium proceedings]. Philadelphia: Chemical Heritage Foundation, 1997.

**Louis Galambos, with Jane Eliot Sewell.** *Networks of Innovation: Vaccine Development at Merck, Sharpe & Dohme, and Mulford, 1895–1995.* New York: Cambridge University Press, 1995.

**Arthur Kornberg.** *The Golden Helix: Inside Biotech Ventures.* Sausalito, Calif.: University Science Books, 1995.

**Fu-Kuen Lin et al.** "Cloning and Expression of the Human Erythropoietin Gene." *Proceedings of the National Academy of Sciences* 82 (Nov. 1985), 7580–7584.

**Meirion B. Llewelyn; Robert E. Hawkins; Stephen J. Russell.** "Discovery of Antibodies." *British Medical Journal* 305:6864 (21 Nov. 1992), 1269–1272.

**César Milstein.** "Nobel Lecture, 8 December 1984. From the Structure of Antibodies to the Diversification of the Immune Response." *Scandinavian Journal of Immunology* 37:4 (April 1993), 385–398.

**William Rutter.** "The Department of Biochemistry and the Molecular Approach to Biomedicine at the University of California, San Francisco." An oral history conducted in 1992 by Sally Smith Hughes, Ph.D., Regional Oral History Office, The Bancroft Library, University of California, Berkeley, 1998.

**Susan Wright.** "Recombinant DNA Technology and Its Social Transformation, 1972–1982." *Osiris* 2 (1986), 303–360.

15 AND 16. FINDING TOMORROW'S MEDICINES

Alliance for the Prudent Use of Antibiotics. www.healthsci.tufts. edu/apua/apua.html.

**Geoffrey Carr.** "From Blunderbuss to Magic Bullet" [on combinatorial chemistry]. *The Economist* (21 Feb. 1998), 59.

**Marlene E. Haffner.** "The Impact of the Orphan Drug Act." *Modern Drug Discovery* 1 (Sept./Oct. 1998), 45–52.

**Celia M. Henry.** "The Next Pharmaceutical Century." In *The Pharmaceutical Century: Ten Decades of Drug Discovery*, pp. 239–242, 244, 247–248, 251. Washington, D.C.: American Chemical Society, 2000.

**Robert Langer et al.** "Biodegradable Long-Circulating Polymeric Nanospheres." *Science* 263 (1994), 1600–1603.

**Mark S. Lesney; Jennifer B. Miller.** "Harnessing Genes, Recasting Flesh." In *The Pharmaceutical Century: Ten Decades of Drug Discovery*, pp. 149–151, 154, 156, 159–160, 163. Washington, D.C.: American Chemical Society, 2000.

**Patricia B. Lieberman; Mary G. Wootan.** "Protecting the Crown Jewels of Medicine" [on antibiotics]. Washington, D.C.: Center for Science in the Public Interest, 1998. www.cspinet. org/reports/abiotic.htm.

**John S. Patton.** "Deep-Lung Delivery of Proteins." *Modern Drug Discovery* 1 (Sept./Oct. 1998), 19–28.

**Mark H. Richmond.** *Human Genomics: Prospects for Health Care and Public Policy.* London: Pharmaceutical Partners for Better Healthcare, 1999.

**Ronald S. Rogers.** "Sepracor: Skating on 'ICE.'" *Chemical and Engineering News* 76:48 (30 Nov. 1998), 11–13.

**John T. Santini; Michael J. Cima; Robert Langer.** "A Controlled-Release Microchip." *Nature* 397 (1999), 335–338.

**Bob Sinclair.** "Everything's Great When It Sits on a Chip: A Bright Future for DNA Arrays." *The Scientist* 13:11 (24 May 1999), 18–20.

**Leonard T. Skeggs, Jr.** "Persistence . . . and Prayer: From the Artificial Kidney to the AutoAnalyzer." *Clinical Chemistry* 46:9 (2000), 1425–1436.

**Wayne Smith.** "Mapping and Sequencing the Human Genome: A Beginner's Guide to the Computational Science Perspective." New York: ACM [Association for Computing Machinery], 2000. www.acm.org/crossroads/xrds4-1/genome. html.

**Andrew Streitwieser.** "History of Computational Chemistry: A Personal View." In *Encyclopedia of Computational Chemistry*, vol. 2, pp. 1237–1244. New York: Wiley, 1998.

**Wendy A. Warr.** "Combinatorial Chemistry." In *Encyclopedia of Computational Chemistry*, vol. 1, pp. 407–417. New York: Wiley, 1998.

**Alan R. Williamson; Keith O. Elliston; Jeffrey L. Sturchio.** "The Merck Gene Index, A Public Resource for Genomics Research." *Journal of NIH Research* 7 (1 Aug. 1995), 61–63.

**Manfred E. Wolff,** ed. *Burger's Medicinal Chemistry and Drug Discovery.* Fifth edition. New York: Wiley, 1995. See John A. Montgomery, "Current and Future Trends in Medicinal Chemistry"; George A. Condouris, "Pharmacology: Current and Future Trends"; B. Veerapandian, "Three-Dimensional Structure-Aided Drug Design"; and George Zografi, "Recent Trends and the Future of Pharmaceuticals."

## General Histories and Overviews

**Alfred Burger.** *Drugs and People: Medications, Their History and Origins, and the Way They Act.* Revised edition. Charlottesville: University of Virginia, 1988.

**Joseph S. Fruton.** *A Bio-Bibliography for the History of the Biochemical Sciences since 1800.* Second edition. Philadelphia: American Philosophical Society, 1992.

———. *Molecules and Life: Historical Essays on the Interplay of Chemistry and Biology.* New York: Wiley-Interscience, 1972.

**Gregory J. Higby; Elaine C. Stroud,** eds. *The Inside Story of Medicines: A Symposium.* Madison, Wis.: American Institute of the History of Pharmacy, 1997.

**Ralph Landau; Basil Achilladelis; Alexander Scriabine,** eds. *Pharmaceutical Innovation: Revolutionizing Human Health.* Philadelphia: Chemical Heritage Foundation, 1999.

**John Mann.** *The Elusive Magic Bullet: The Search for the Perfect Drug.* New York: Oxford University Press, 1999.

*The Pharmaceutical Century: Ten Decades of Drug Discovery.* Washington, D.C.: American Chemical Society, 2000.

*Pharmazie.* Munich: Deutsches Museum, 2000.

**Walter Sneader.** *Drug Prototypes and Their Exploitation.* New York: Wiley, 1996.

**Mikuláš Teich, with Dorothy M. Needham.** *A Documentary History of Biochemistry: 1770–1940.* Rutherford, N.J.: Fairleigh Dickinson University Press, 1992.

## Collected Biographies

*American Chemists and Chemical Engineers.* Edited by Wyndham D. Miles. Washington, D.C.: American Chemical Society, 1976. See articles on John Jacob Abel, Edward Adelbert Doisy, Philip Hench, Edward Calvin Kendall, Tadeus Reichstein, Edward Robinson Squibb, Albert Szent-Györgyi, Max Tishler, Harvey Washington Wiley, and Robert Burns Woodward.

*American National Biography.* Edited by John A. Garraty and Mark C. Carnes. New York: Oxford University Press, 1999. See articles on John Jacob Abel, Charles Best, Rachel Fuller Brown, Walter Campbell, Edward Adelbert Doisy, Philip Hench, Percy Lavon Julian, Edward Calvin Kendall, Karl Paul Gerhardt Link, Leonor Michaelis, Albert Sabin, Jonas Salk, Edward Robinson Squibb, Albert Szent-Györgyi, Jokichi Takamine, Max Tishler, Selman Abraham Waksman, Harvey Washington Wiley, Robert Ramapatnam Williams, and Robert Burns Woodward.

*The Biographical Dictionary of Scientists.* Edited by Roy Porter. New York: Oxford University Press, 1994. See articles on Frederick Grant Banting, Paul Berg, Ernst Boris Chain, Henry Hallett Dale, Gerhard Domagk, Paul Ehrlich, Alexander Fleming, Howard Walter Florey, George Herbert Hitchings, Dorothy Crowfoot Hodgkin, Edward Calvin Kendall, César Milstein, Louis Pasteur, Pierre-Joseph Pelletier, Albert Bruce Sabin, Jonas Edward Salk, Albert Szent-Györgyi, and Robert Burns Woodward.

*Biographical Memoirs of Fellows of the Royal Society.* London: Royal Society, 1955–. See articles on Charles Best (Vol. 28), Daniel Bovet (Vol. 39), Ernst Chain (Vol. 29), James Collip (Vol. 19), Henry Dale (Vol. 16), Gerhard Domagk (Vol. 10), Alexander Fleming (Vol. 2), Howard Florey (Vol. 17), Austin Hill (Vol. 4), Otto Loewi (Vol. 8), Tadeus Reichstein (Vol. 45), and Robert Woodward (Vol. 27).

*Biographical Memoirs of the National Academy of Sciences, 1877–.* Washington, D.C.: National Academy Press, 1931. See articles on John J. Abel (Vol. 24), Gertrude B. Elion (Vol. 78), Percy L. Julian (Vol. 52), Edward C. Kendall (Vol. 47), Karl P. Link (Vol. 65), Leonor Michaelis (Vol. 31), Max Tishler (Vol. 66), and Robert Burns Woodward (Vol. 80).

*Dictionary of Scientific Biography.* Edited by Charles Coulston Gillespie. New York: Scribners, 1970–1990. See articles on John Jacob Abel, Frederick Banting, Charles Herbert Best, Ernst Boris Chain, James Collip, Henry Hallett Dale, Carl Peter Henrik Dam, Edward Adelbert Doisy, Gerhard Domagk, Paul Ehrlich, Alexander Fleming, Howard Florey, Percy Lavon Julian, Otto Loewi, Leonor Michaelis, Louis Pasteur, and Selman Abraham Waksman.

The Nobel Foundation. Available at http://www.nobel.se (cited 30 September 2002). See articles on Frederick Grant Banting, Paul Berg, Sir James W. Black, Daniel Bovet, Ernst Boris Chain, Stanley Cohen, Sir Henry Hallett Dale, Henrik Carl Peter Dam, Edward Adelbert Doisy, Gerhard Domagk, Paul Ehrlich, Gertrude B. Elion, Sir Alexander Fleming, Sir Howard Walter Florey, Philip Showalter Hench, George H. Hitchings, Dorothy Crowfoot Hodgkin, Edward Calvin Kendall, Georges J. F. Köhler, Otto Loewi, John James Richard Macleod, César Milstein, Tadeus Reichstein, Albert von Szent-Györgyi Nagyrapolt, Selman Abraham Waksman, and Robert Burns Woodward.

*Nobel Laureates in Medicine or Physiology.* Edited by Daniel M. Fox, Marcia Meldrum, and Ira Rezak. New York: Garland Publishing, 1990. See articles on Frederick Grant Banting, James Whyte Black, Daniel Bovet, Ernst Boris Chain, Stanley Cohen, Henry Hallett Dale, Carl Dam, Edward Adelbert Doisy, Gerhard Domagk, Paul Ehrlich, Gertrude Belle Elion, Alexander Fleming, Howard Walter Florey, Philip Hench, George Hitchings, Edward Calvin Kendall, Georges Köhler, Otto Loewi, John Macleod, Leonor Michaelis, César Milstein, Louis Pasteur, Tadeus Reichstein, Albert Szent-Györgyi, and Selman Abraham Waksman.

*Notable Twentieth-Century Scientists.* Edited by Emily J. McMurray. New York: Gale Research, 1995. See articles on John Jacob Abel, Alfred W. Alberts, Frederick Banting, Paul Berg, Charles Herbert Best, James Whyte Black, Daniel Bovet, Herbert W. Boyer, Rachel Fuller Brown, Ernst Boris Chain, Stanley N. Cohen, James Bertram Collip, Henry Hallett Dale, Carl Djerassi, Edward Doisy, Gerhard Domagk, Paul Ehrlich, Gertrude Belle Elion, Gladys Anderson Emerson, Alexander Fleming, Howard Walter Florey, Elizabeth Lee Hazen, Philip Hench, George Hitchings, Gladys Lounsbury Hobby, Dorothy Crowfoot Hodgkin, Percy Lavon Julian, Edward Kendall, Georges Köhler, Otto Loewi, John Macleod, Leonor Michaelis, César Milstein, Louis Pasteur, Tadeus Reichstein, Albert Sabin, Jonas Salk, Albert Szent-Györgyi, Max Tishler, Selman Abraham Waksman, and Robert Burns Woodward.

*Obituary Notices of the Royal Society.* London: Royal Society, 1932–1954. See articles on John Abel (Vol. 2), Frederick Banting (Vol. 4), and John Macleod (Vol. 1).

*Proceedings of the Royal Society of London.* London: Royal Society, 1830–1904. See article on Louis Pasteur (Vol. 62).

*Proceedings of the Royal Society of London Series B.* London: Royal Society, 1905–1931. See article on Paul Ehrlich (Vol. 92).

**Marelene Rayner-Canham; Geoffrey Rayner-Canham.** *Women in Chemistry: Their Changing Roles from Alchemical Times to the Mid-Twentieth Century.* Philadelphia: Chemical Heritage Foundation, 2001. See articles on Rachel Fuller Brown, Gertrude Elion, Elizabeth Hazen, Dorothy Hodgkin, and Maud Menten.

*Women in Chemistry and Physics: A Biobibliographic Sourcebook.* Edited by Louise S. Grinstein, Rose K. Rose, and Miriam H. Rafailovich. Westport, Conn.: Greenwood Press, 1993. See articles on Gertrude Elion, Gladys Emerson, and Dorothy Hodgkin.

ORAL HISTORIES

(All of the following oral histories are on deposit at the Othmer Library of the Chemical Heritage Foundation. Those with limited access are so noted.)

**Albert Carr.** Oral history interview by James G. Traynham at Hoechst-Marion Roussel Center, 12 November 1999. Limited access.

**Carl Djerassi.** Oral history interview by Jeffrey Sturchio and Arnold Thackray at Stanford University, 31 July 1985. Limited access.

**Russell Marker.** Oral history interview by Jeffrey Sturchio at Pennsylvania State University, 17 April 1987.

**Robert McNeil.** Oral history interview by Mary Ellen Bowden and Arnold Thackray in Philadelphia, Pennsylvania, 13 and 30 August 2001, and Wyndmoor, Pennsylvania, 15 August 2002. Limited access.

**Miguel Ondetti.** Oral history interview by James J. Bohning in Princeton, New Jersey, 12 January 1995.

**George Rathmann.** Oral history interview by Arnold Thackray, Leo Slater, and David Brock in Philadelphia, Pennsylvania, 16 and 17 September 1999. In process.

**Lewis Sarett.** Oral history interview by Leon Gortler in Viola, Idaho, 6 September 1990.

**Max Tishler.** Oral history interview by Leon Gortler and John Heitmann at Wesleyan University, 14 November 1983.

# Index

3M Company, 187
6-mercaptopurine (6-MP). *See* Purinethol

Abbott Laboratories, 187
Abel, John Jacob, 47–51
Abraham, Edward Penley, 78
acetaminophen, 16
acetanilide. *See* Antifebrin
acetic acid, 16
acetylcholine, 61, 63, 68, 70, 164, 168
acetylsalicylic acid. *See* aspirin
actin, 32
actinomycetes, 78, 90, 92, 93
actinomycin, 92
acyclovir. *See* Zovirax
Addison's disease, 97, 103
adenine, 172
adrenal glands, 97, 99, 101
Adrenalin, 97
adrenaline, 47, 49, 61, 68, 70, 163, 164. *See also* epinephrine
adulteration, of food and drugs, 20, 21, 23, 25, 26, 28, 136
agranulocytosis, 150
Ahlquist, Raymond, 148
AIDS, treatment of, xi, 42, 130, 169–177, 184
Albers-Schönberg, Georg, 157–162
Alberts, Alfred W., 145, 157–162
Aldrich, Thomas, 49
Allegra, 164, 166
allergies, symptoms of, 164, 165; treatment of, 164–166
Allison, V. D., 73
allopurinol. *See* Zyloprim
American Medical Association, conventions of 1876 and 1877, 25
American Society for the Prevention of the Adulteration of Food, 26
Ames Company, 14
Amgen (Applied Molecular Genetics), 187, 190
amphetamines, 49, 113
ampicillin, 81
amylase, 51
anemia, treatment of, 187, 190
aneurin, 33
angiotensin, 153, 154, 156
angiotensin-converting enzyme (ACE) inhibitors, 151, 153, 154, 161, 175
angiotensinogen, 151, 153

Antergan, 165
anthrax, 8, 9
antibiosis, 73
antibiotics, x, xi, 19, 66, 72–95, 178; forerunners of, 64; future research on, 198–199; resistance to, 199
antibodies, 17, 185
anticholesterol drugs, 157, 160, 161, 162
antidepressant drugs, xi, 163–168
Antifebrin, 13, 14
antifungal medicines, 93
antigens, 185, 187, 196, 196
antihistamines, xi, 42, 118, 163–168; nonsedating, 164–166
antipsychotic drugs, 113–124, 163, 164
antisepsis, 73
antitoxins, 9, 17, 95
antiulcer drugs, xi, 146–150
artificial kidney (early dialysis machine), 51
ascorbic acid, 30, 32
aspirin (acetylsalicylic acid), 2, 10, 12, 153
Association of Official Agricultural Chemists, 26
Atabrine, 65
attenuation, 8, 9, 125, 133, 173
Attenuvax, 131
Aureomycin, 115
Aventis Pasteur, 9
Axelrod, Julius, 14, 16
azathioprine. *See* Imuran
azidothymidine (AZT), 172, 173, 177, 200
azo dyes, 65

Bache, Benjamin Franklin, 23
bacterial virus lambda, 180–181, 182, 183
bacteriophage, 190
Bactrim, 64
Baekeland, Leo, 51
Baeyer, Adolf von, 10
Banting, Frederick Grant, 1, 51, 52–56, 57
barbiturates, 113, 114
Barger, George, 59, 61
Baumann, Eugen, 47
Bayer Company, 10, 12, 14, 51, 65
Bayliss, William Maddock, 59
Beecham Group, 81
Behring, Emil von, 17
Bell Laboratories, 33

Benzedrine, 49, 113
benzoheptoxdiazines, 122
Berg, Paul, 179–183
beriberi, 29, 33
Bernal, John Desmond, 82
Bertheim, Alfred, 19
Berzelius, Jöns Jakob, 43
Best, Charles Herbert, 1, 52–56, 57
beta-blockers, 148, 150
beta-lactam ring structure, for penicillin, 81, 115
bile acids, 96
bioassays, 205
Biogen, 191, 193
biotechnology, 178–206
biotin, 122
Black, James Whyte, 147, 148–150, 151, 163, 172
blood pressure–lowering drugs, 150, 154, 156
Blumberg, Baruch, 194
Bodanszky, Miklos, 151
Botanica-Mex, 110
Bovet, Daniel, 68–70, 81, 118, 147, 163, 164
Bovet Nitti, Filomena, 68, 70
Boyer, Herbert W., 179–183
bradykinin, 151, 153
Bristol-Myers Squibb, 20, 41, 154, 156
Brodie, Bernard, 14, 16
bronchodilators, 148
Brooklyn Naval Hospital, 23
Brown, Michael, 157
Brown, Rachel Fuller, 93–95
Buchner, Eduard, 43
burimamide, 150
Burnet, Frank Macfarlane, 133
Burroughs Wellcome and Company, 170
Butenandt, Adolf, 35, 109
Byers, L. D., 153

caffeine, 3
Cahn, Arnold, 13
Campbell, Walter, 134, 136–138
cancer, 169–177; of the brain, treatment of, 200; of the lung, 141; treatment of, 184
cancer drugs, xi, 169–177
Cantell, Kari, 191
captopril, 153
cardiovascular diseases, treatment of, 42, 146–162

cardiovascular drugs, xi, 146–162
Carini, David, 154–156
Carr, Albert A., 164–166
Cassella (chemical company), 19
Caventou, Joseph-Bienaimé, 3–5
Cetus, 191
Chain, Ernst Boris, 73, 76–81, 82, 84
Chang, Annie, 183
Charpentier, Paul, 70, 118–121, 163
Chemie Grünenthal, 142
Chen, Julie, 161
Chen, Ko Kuei, 49
chicken leukemia, 131
chiral center, 117
Chiron Corporation, 196
Chiu, Andrew, 154
chlordiazepoxide. See Librium
chloroamphetamine, 168
chlorophyll, 3, 114
chlorpromazine. See Thorazine
cholecystolinin, 151
cholesterol, 96, 157, 160, 161, 162;
    synthesis of, 157, 160, 161
cholestyramine, 160
CIBA, 110
CIBA-Geigy, 115, 117
cigarette smoking, 141
cimetidine. See Tagamet
Civil War, 21, 26
Claritin, 164
clinical testing, of drugs, 134–144
cloning, 183, 187, 191, 193, 194
Clowes, George Henry Alexander, 49,
    57–58
codeine, 12, 16
Coghill, Robert D., 86
Cohen, Stanley N., 179–183
colchicine, 117
Collip, James Bertram, 52–56
colorimetric analysis, 204, 205
Colton, Frank B., 112
combinatorial chemistry, 197, 203–205
compactin, 161, 162
compound A (11-dehydrocorticosterone),
    99, 101
compound E, 99, 101, 103, 106. See also
    cortisone
Compton, Walter Ames, 14
computational chemistry, 197, 205
Condra, Jon, 173–177
Conklin, Raymond L., 14
Consumer's Research, 136

Contergan, 142
contraceptives, oral, xi, 112
Copeland, Royal, 136
Cornely, Donald A., 16
cortical hormones, 96, 99
cortin, 97
cortisone, 96, 97, 103, 105, 106, 111,
    115. See also compound E
Coumadin, 38, 41
Council on Pharmacy and Chemistry
    (1905), 28
Courvoisier, Simone, 118–121
cowpox virus, 125
Cozaar, 154, 156
Crixivan, 173, 177
curare, 70
Cushing's disease, 103
Cushman, David, 153
Cutter Laboratories, 59–63, 90, 147, 164

Dale, Henry Hallett, 59–63
Dalldorf, Gilbert, 93, 95
Dam, Carl Peter Hendrik, 38–41
Daniels, Marc, 141
Dawson, Martin Henry, 84
Deficiency Appropriations Act of 1919,
    136
Delay, Jean, 118–121
Deniker, Pierre, 118–121
deoxyribonucleic acid (DNA), 169, 172,
    173, 175, 178, 180, 181, 183, 190,
    191, 193, 196, 201, 205
depression, treatment of, 167–168
desoxycholic acid, 99
diabetes mellitus, treatment of, 42, 52,
    54–56, 57
diaminopurine, 172
diastase, 51
diazepam. See Valium
dichloroisoproterenol, 148
dicumarol, 41
Diels-Alder reaction, 117
diethylene glycol, 138, 142
digitalis (foxglove), 147
digitoxin, 146
diosgenin, 109, 110
Diosynth, 110
diphenhydramine, 118
diphtheria, 17
Djerassi, Carl, 106–112
DNA. See deoxyribonucleic acid
DNA chips, 203, 205

DNAX, 183
Dobbin, James C., 23
Doisy, Edward Adelbert, 38–41
Doll, Richard, 141
Domagk, Gerhard, 1, 65–67, 68
Donath, Willem, 33
dopamine, 164, 168, 200
Dorsey, Bruce, 173–177
double-blinded clinical trial, 139. See also
    clinical testing, of drugs
Dreser, Heinrich, 12
drug development process, new, 135
drugs, delivery of, 200; deep-lung, 201
Dubos, René, 78, 90
Dudley, H. W., 63
Dulbecco, Renato, 180
Dumas, Jean-Baptiste, 3
Duncia, John, 154–156
DuPont Merck Pharmaceutical Com-
    pany, 156
DuPont Pharmaceuticals, 41, 154, 156
Durant, Graham, 150

E. R. Squibb and Company, 20
E. R. Squibb and Sons, 131, 86, 95, 151
ear infection, treatment of, 64
Eco RI (enzyme), 180–181, 182
Eggers Doering, William von, 115
Ehrlich, Paul, x, 2, 17–19, 43, 46, 59, 65,
    68, 73, 90, 147, 203
Eijkman, Christiaan, 33
Eingruen, Arthur, 12
Eli Lilly and Company, 49, 57, 127, 148,
    167, 168, 183
Elion, Gertrude, 145, 147, 151, 169,
    170–172
Elliott, T. R., 59
Emerson, Gladys Ludwina Anderson, 1,
    35–37
emetine, 3
Emini, Emilio, 173–177
Emmett, John, 150
enalapril. See Vasotec
encephalitis, Japanese B, 131
Enders, John, 126, 127
Endo, Akira, 161
Endo Products, Inc., 41
endocarditis, subacute bacterial, 84
enzymes, 42–46, 51
ephedrine, 49
epinephrine, xi, 47, 49, 51, 148. See also
    adrenaline

Epogen, 190
ergot, 59
ergotoxine, 59, 61
erythromycin, 166
erythropoietin, 178, 187, 190, 200; recombinant human, 187, 190
Eschenmoser, Albert, 117
*Escherichia coli*, 181, 182, 183
estradiol, 110
Ethyl Gasoline Corporation, 109
Euler, Ulf von, 167
Evans, Herbert M., 35
Ewins, Arthur, 61

FDA. *See* U.S. Food and Drug Administration
Feldman, William H., 92
fermentation, 6, 91; of penicillin, 81, 86, 89; submerged (or deep-tank), 86, 89; surface, 89
fermentation products for screening (FERPS) system, 161
"ferments," 43
fertility, and vitamin-E deficiency, 35, 37
fexofenadine. *See* Allegra
Fischer, Emil, 17, 43, 147
Fisher, Ronald Aylmer, 139
Fleming, Alexander, 71, 73–75, 85
Flexner, Simon, 109
Flinn, Frederick B., 14, 16
Florey, Howard Walter, 73, 76–81, 84, 86
fluoxetine. *See* Prozac
Folkers, Karl, 157, 161
Food, Drug, and Cosmetic Act of 1938, 14, 134, 136, 138, 144
Food, Drug, and Insecticide Administration, 136
Fourneau, Ernest, 68
fowl cholera, 6, 8
Francis, Thomas, Jr., 126, 127
Fränkel, Sigmund, 122
free radicals, 46
Friedheim, Ernst, 46
Fröhlich, Alfred, 30
Fukui, Kenichi, 117
Fuller, Ray W., 167–168
Fulton, John, 80
fungal infections, 93
fungi, 90, 93, 95
fungicidin, 95
Funk, Casimir, 29

G. D. Searle and Company, 112

galactosemia, 194
Ganellin, Robin, 150
Geismar, Lee, 142
Gene Chip probe array, 203
gene therapy, 198
Genentech, 183, 191, 196
George K. Smith and Company, 20
Georg-Speyer-Haus, 19
Gerhardt, Charles, 12
Gilbert, Walter, 181
glaucoma, treatment of, 63, 106
GlaxoSmithKline, 20, 21, 170
Glidden Company, 106
glycosuria, 54
Goldstein, Joseph, 157
Goldwasser, Eugene, 187, 190
Gonzalez, Arelina, 111
Gould Amendment of 1913, 136
gout, treatment of, 117, 170, 172
Greenberg, Leon, 14, 16
Greenwood, Major, 139
guanine, 169, 172

H. K. Mulford Company, 49
Haldane, J. B. S., 46
Hall, Benjamin D., 196
haloperidol, 164, 165
Hansen, Gaurth, 194
Harington, Charles, 76
Harrer, Paul, 32
Harrington, C. R., 97
Hata, Sahashiro, 19, 90
Haworth, Norman, 101
Haworth, Walter, 30, 32
Hazen, Elizabeth Lee, 93–95
Heatley, Norman, 78, 80, 86
Helling, Robert, 182
hemopoietine, 187
Hench, Philip Showalter, 97–105
Henshaw, H. Corwin, 92
hepatitis, treatment of, 193, 194, 196
hepatitis B, vaccine for, 131, 187, 194, 196
Hepp, Paul, 13
Heptavax-B, 133, 194, 196
heroin, 10, 12
herpesvirus, treatment of, 170, 172
hexuronic acid, 30
Heyden Company, 10
high-throughput screening (HTS), 197, 205, 206
Hill, Austin Bradford, 134, 139–141
Hilleman, Maurice, 131–133, 196

Hinsberg, Oscar, 14
histamine-2 blockers, 148, 150
histamines, 59, 61, 163, 164–165
Hitchings, George, 145, 147, 151, 169, 170–172
Hobby, Gladys L., 71, 84–85
Hodgkin, Dorothy Crowfoot, 56, 71, 81, 82–83, 115
Hoechst (chemical company), 19
Hoffman, Carl, 161
Hoffmann, Felix, 10–12, 17
Hoffmann–La Roche, 122, 123, 175
Hoffmann, Roald, 117
Hogan, John, 16
Holst, Alex, 30
Hopkins, Frederic Gowland, 76
Hörlein, Heinrich, 65
hormone replacement therapy, 96
hormone therapies, xi
hormones, 42, 47–58
HTS. *See* high-throughput screening
Huff, Joel, 173–177
human immunodeficiency virus (HIV), treatment of, 173, 175, 177
human insulin, 183, 198
Humulin (human insulin), 183
hybridoma technology, xii, 178, 184, 185, 186
hydrocortisone, 106
hydroxymethylglutaryl coenzyme A (HMG-CoA) reductase, 161
hypertension, treatment of, 151, 153, 154, 156

IG Farbenindustrie, 65
Iletin (insulin), 57
imipramine. *See* Tofranil
immune gamma globulin, 131, 133
immune response, 17, 19
immunoglobulins, 185
Imperial Chemical Industries (ICI), 80, 148
Imuran, 172
indinavir sulfate. *See* Crixivan
influenza, vaccine for, 126, 131
influenza virus, 126
infrared spectroscopy, 205
Institute for Infectious Diseases (Berlin), 17
Institute for Serum Research and Serum Testing (Berlin), 17
Institute for the Study of Analgesic and Sedative Drugs, 14

insulin, xi, 42, 51, 52, 54–56, 57, 63, 82, 151, 178
interferon, 191, 193
ipecac, 3
Isaacs, Alick, 191
islets of Langerhans, 52, 54
isoniazid, 66
Itakura, Keiichi, 183
ivermectin, 162

Jansen, Barend, 33
jaundice, 97
Jenkins, Thomas E., 23
Jenner, Edward, 8, 125
Jiro, Oyam, 142
John K. Smith and Company, 21
Johnson & Johnson, 16
Julian Laboratories, 106
Julian, Percy, 71, 106–112

Kade, Charles F., Jr., 16
Kaiser, Dale, 181
Kalckar, Herman, 179
Kalle and Company, 13
kanamycin, 183
Kaposi's sarcoma, treatment of, 193
Kefauver-Harris Amendments of 1962, 134, 141, 142, 144, 147
Kelsey, Frances Oldham, 71, 134, 142–144
Kendall, Edward Calvin, 97–105
Kennedy, John F., 71, 143, 144
ketoconazole, 166
King, Charles G., 30, 32
Kitasato, Shibasaburo, 17
Klarer, Josef, 65
Koch, Robert, 17
Köhler, Georges, 184–186
Kolbe, Hermann, 10
Kornberg, Arthur, 179–180
Kraut, Carl J., 12
Kühne, Willy, 43

Laborit, Henri-Marie, 118–121
Laidlaw, P. P., 61
Lamont, David S., 16
Langerhans, Paul, 52
lanosterol, 161
Largactil, 119
Lasker, Mary, 92, 133
LDLs. See low-density lipoproteins
L-dopa, 200
Lederle Laboratories, 86

Lehmann, Frederico, 110
Lescohier, Alexander, 110
Lester, David, 14, 16
leukemia: hairy-cell, treatment of, 193; treatment of, 170, 172
Librium, 113, 122, 123
Lin, Fu-Kuen, 145, 187–190
Lind, James, 30, 33
Lindenmann, Jean, 191
Link, Karl Paul Gerhardt, 38–41
liquid chromatography, 205
Lobhan, Peter, 181
Loewi, Otto, 59–63, 70
London School of Hygiene and Tropical Medicine (LSHTM), 139
losartan. See Cozaar
Louisville Chemical Works, 23
lovastatin. See Mevacor
Lovell, Douglas G., Jr., 16
low-density lipoproteins (LDLs), 157, 160
lysergic acid diethylamide (LSD), 115
lysis, 75
lysozyme, 73, 78

Macleod, John James Rickard, 51, 52–56, 57
Magendie, François, 3
malaria, 2; treatment of, 2, 3, 21, 65, 68, 115, 170, 172
MAOIs. See monoamine oxidase inhibitors
Marek's disease, 131
Marine Biological Laboratories (Woods Hole, Mass.), 57
Marker, Russell, 106–112
Marsh, David, 16
mass spectroscopy, 112, 161, 162, 205
Mayo Clinic, 97, 101, 103
McKeen, John Elmer, 88–89
McNary-Mapes Amendment (1930), 136
McNeil, Henry, 16
McNeil Laboratories, New Products Committee, 13, 16
McNeil, Robert L., Jr., 13–16
measles, killed- and live-virus vaccines for, 131
measles, mumps, and rubella (MMR), vaccine for, 125
Medical Research Council (MRC), 139, 141
melanoma, malignant, treatment of, 193

meningitis, treatment of, 172
Menten, Maud Leonora, 43–46
meprobamate. See Miltown
Merck and Company, 33, 86, 92, 99, 101, 103, 105, 115, 157, 160, 161, 162, 173, 175, 177, 194, 196
Merck Institute for Therapeutic Research, 37, 131
Merck, Sharp and Dohme Research Laboratories, 105
Mering, Joseph von, 14, 52
Merrell's "library" of molecules, 165–166
Merrifield, Bruce, 202, 203
messenger RNA (mRNA), 191, 193
Metchnikoff, Élie, 9, 17
methemoglobinemia, 14
metiamide, 150
Mevacor, 157, 162
mevalonic acid, 157, 161, 162
mevastatin, 161
Mexican-American War, 21, 23
Meyer, Karl, 84
Michaelis, Leonor, 43–46
microchip, silicon, for drug delivery, 201
Mietzsch, Fritz, 65
Miles Laboratories, 14
Milstein, César, 184–186
Miltown, 122
Minkowski, Oskar, 52
MMR. See measles, mumps, and rubella, vaccine for
Molloy, Bryan B., 167–168
Monaghan, Richard, 161
monoamine oxidase inhibitors (MAOIs), 167
monoclonal antibodies, 178, 184, 186
Monsanto, 115
morphine, 3
Moyer, Andrew J., 86–87
MRC. See Medical Research Council
mumps, vaccine for, 131; killed-virus, 131, 133
Mumpsvax, 133
mycostatin, 93, 95
myelomas, 185
myocardial infarction, treatment of, 41
myosin, 32

Nagai, Nagajosi, 49
nanospheres, synthetic polymer, 200
naphthalene, 13
National Foundation for Infantile Paralysis, 127

National Institute of Medical Research (NIMR), 61, 63
National Institutes of Health, 181
National Pure Food and Drug Congress of 1898, 26
National Research Council (NRC), 99, 101
Navratil, Ernst, 61
Nazi Party, 66
Neosalvarsan, 19
nerve impulses, chemical transmission of, 59, 61, 63
neurotransmitters, 42, 59–63, 164, 167, 168
new drug application (NDA), 134, 135
New York State Medical Society, convention of 1879, 25
NIMR. See National Institute of Medical Research
nitroglycerin, 146
Nitti, Federico, 68
Nitti, Francesco Saverio, 68
NMR. See nuclear magnetic resonance
norepinephrine (noradrenaline), 148, 164, 167, 168
norethindrone, 112
Nourse, Walter B., 95
Noyes, Arthur A., 51
NRC. See National Research Council
nuclear magnetic resonance (NMR), 161, 162, 205
nystatin. See mycostatin

O'Brien, W. J., 106
Ochoa, Severo, 191
O'Connor, Basil, 126
oligonucleotides, 190
Oliver, George, 47
Ondetti, Miguel, 145, 151–153, 175
Opie, Eugene, 52
optical rotatory activity, 6, 200
optical rotatory dispersion, 112
Orphan Drug Act of 1983, 199
orphan drugs, 199–200
osmium tetraoxide, 103
oxazolone structure, for penicillin, 81

paprika, 30, 32
Parke, Davis and Company, 49, 51, 59, 97, 109, 110, 127
Parkinson's disease, treatment of, 200
Parsons, Michael, 150
Pasteur, Louis, x, 2, 6–9, 125, 170

Patchett, Arthur A., 145, 157–162
Pauli, Wolfgang, 122
Pediazole, 64
Pelletier, Pierre-Joseph, 3–5
Penhoet, Edward, 196
penicillin, 66, 72, 73, 75, 76, 78, 80, 81, 82, 84, 86, 89, 92, 115, 178; testing on humans, 78, 80
penicillin V, 81
pepsin, 82
peptide synthesizer, 202, 203
peptides, 151
peripheral neuritis, with thalidomide use, 142
Perón, Juan, 151
peroxidase, 30
Pfizer and Company, 49, 84, 86, 89, 115, 130
pharmacology, 47
phenacetin, 14
phenobarbital, 122
phenothiazine, 118
phocomelia, with thalidomide use, 142, 144
physostigmine, 63, 106
Piaget, Jean, 68
Pikl, Josef, 106
placebo, 139
plasmids, 181, 182, 183, 193, 196
pneumonia, 95
Pneumovax, 133
Polaroid Corporation, 115
poliomyelitis, vaccine for, 125, 126, 127; killed-virus, 126, 127, 130; live-virus, 127, 130
polioviruses, 126–127
polysaccharides, 95
Powell, C. E., 148
pregnanediol, 109
Premarin (hormone replacement therapy), 96
Pritchard, Brian, 150
progesterone, 96, 106, 109, 110, 111, 112
promethazine, 118
pronethalol, 148, 150
Prontosil, 66, 68
propranolol, 150
Prosympal, 68
protease inhibitors, 175
Prozac, 113, 167, 168
pSC101 (plasmid), 183
ptyalin, 51
puerperal fever, treatment of, 78

Pugwash movement, 82
Pure Food and Drugs Act of 1906, xi, 20, 21, 26, 28, 134, 136
purines, 170, 172
Purinethol, 172
pyocyanase, 78
pyramethamine, 172

quinine, 2, 3, 21, 65, 68, 115

rabies, 6, 8
Randall, Lowell, 122–124
randomization, 139
randomized controlled trial, 139, 141
Rathmann, George, 187–190
rational drug design, xii, 146, 170, 172
recall, of drug, 138
receptor theory, 148
recombinant DNA (rDNA) technology, xii, 178, 179, 181, 183, 187, 190, 191, 194, 196
Recombivax HB, 196
regulation, of food and drug manufacture, 136, 138, 142, 144
Reichstein, Tadeus, 97–105
Reichstein's Substance S, 106, 110
Reid, Roger, 84
renin, 151, 153
replacement therapy, 42
Research Corporation, 34, 95, 117
reserpine, 113, 114, 115, 117
reverse transcriptase inhibitor, 173, 175
rheumatic fever, 84
rheumatoid arthritis, treatment of, 96, 97, 103, 184
Rhodoquine, 68
Rhône Poulenc, 118, 119
riboflavin, 103, 122
ribonucleic acid (RNA), 172, 173, 175, 180, 191
Richards, Alfred Newton, 61, 76, 80, 92
Richards, Ellen Swallow, 47
Ritter, Joseph A., 16
river blindness, treatment of, 162
RNA. See ribonucleic acid
Robbins, Frederick, 126, 127
Robinson, Robert, 80, 81
Rona, Peter, 46
Roosevelt, Franklin Delano, 126, 136, 138
Roosevelt, Theodore, 28
Rosenkranz, George, 110
Roth, James, 16

Roussel-Uclaf, 117
Royal Prussian Institute for Experimental Therapy (Frankfurt), 17
rubella, vaccine for, killed-virus, 133
Rubeovax, 131
Rubin, Bernard, 153
Rutter, William J., 194–196
Ruzicka, Leopold, 101, 109, 122

Sabin, Albert Bruce, 71, 126–130
Sabo, Emily, 153
salicylic acid, 2, 10, 12, 16
Salk Institute for Biological Studies, 130, 180
Salk, Jonas, 126–130
Salser, Winston, 187
salvarsan, 2, 19, 64, 73, 75, 90
Samuel E. Massengill Company, 138
Sandoz Ltd., 117
Sanger, Frederick, 56, 82, 181
Sankyo Pharmaceutical Company, 51
sapogenin, 109
Sarett, Lewis Hasting, 97–105
sarsaparilla root, 109
Schaeffer, Howard, 172
Schäfer, Edward, 47
Schatz, Albert, 92
Schering-Plough Corporation, 193
screening, "virtual," 205. See also high-throughput screening
scurvy, 29, 30, 32, 33
secretin, 59, 151
sedatives, 142
Seldane, 166
selective serotonin reuptake inhibitors (SSRIs), 167, 168
septicemia, treatment of, 172
Septra, 172
serotonin, 163, 164, 167, 168
Sertürner, Friedrich Wilhelm, 3
sex hormones, 96, 106
Shaffer, James M., 16
Sheehan, John, 81
Sherley Amendment of 1911, 136
Shiga, Kiyoshi, 19
Sigal, Irving, 173–177
silkworms, 6
Skeggs, Leonard, 204, 205
Slater, I. H., 148
sleeping sickness, 19
smallpox, 8; vaccine for, 125
Smith, George K., 21–22
Smith, John K., 21

Smith, John L., 89
Smith, Kline and French, 119, 120. See also George K. Smith and Company and John K. Smith and Company
Smith, Kline and French Laboratories, 150
Smith, Laurence, 23
Smith, Mahlon K., 21
SmithKline Beecham, 12, 156
Snyder, Solomon, 168
soil bacteria, 90
somatostatin, 183
Somlo, Emeric, 110
Sørensen, Søren, 46
Squibb, Edward Robinson, 23–25
SSRIs. See selective serotonin reuptake inhibitors
Stadtman, Earl, 160
staphylococcus, and penicillin, 73
Stark, William, 30
Starling, Ernest Henry, 47, 59
Staub, Anne Marie, 70, 164
Stephenson, John, 148
stereoisomers, 200
Sterling Products, 12
Sternbach, Leo, 122–124
steroids, 96–112, 115
stigmasterol, 106
Stokes, Joseph, 131
Strander, Hans, 191
streptococcal infections, 64, 65, 66, 68; hemolytic, 84
streptomycin, 66, 84, 89, 92; clinical trial of, 141
strychnine, 3
succinyl-L-proline, 153
sulfa drugs, xi, 64–71, 172
sulfanilamide, 64, 68; elixir of, 138, 142
sulfonamides, 64, 65, 68, 72
Svirbely, J. L., 30
sweet clover disease, 38–39, 41
Syntex, 110, 111, 112
syntheses, automatic: "solid-phase," 203; "split," 204; "parallel," 204
syphilis, 19, 64, 72, 75
Szent-Györgyi, Albert, 30–32, 76, 97

Tagamet, 150, 198
Takaki, Kanehiro, 33
Takamine Ferment Company, 51
Takamine, Jokichi, 1, 47–51
Takeda Chemical Industries, 154
"Takeda molecules," 154, 156

Taniguchi, Tada, 191
Tannig, 10
teratogens, 142
terfenadine. See Seldane
Terramycin, 84, 89, 115
testosterone, 110
tetanus, 17
tetracycline, 89, 183
tetraethyl lead, 109
thalidomide, 142, 144
thiamine, 33
thioguanine, 172
Thorazine, 70, 113, 114, 118–119, 122, 163, 164, 168
thrombosis, treatment of, 41
thyroxin, 78
thyroxine, 97
Tishler, Max, 97–105, 131
tocopherol, alpha and beta, 35
Tofranil, 113
Tourette's syndrome, 199
tranquilizers, 113–124
Tréfouel, Jacques, 68
Tréfouel, Térèse, 68
tricyclic antidepressants, 167, 168
Trimble, E. L., 21
trimethoprim. See Septra
trypan red, 19
tuberculosis, 17, 72, 92, 139
"Tugwell bill," 136
Tugwell, Rexford Guy, 136, 138
Tylenol, 2, 13, 14, 16

U.S. Department of Agriculture, 136; Chemical Division, 26
U.S. Food and Drug Administration (FDA), 13, 14, 16, 119, 134, 135, 136, 138, 139, 142, 144, 153, 156, 162, 166, 173, 177, 183, 199, 200
U.S. Pharmacopoeia, 1860 Committee for the Revision of, 23
U.S. Pharmacopoeia, 1870 Committee for the Revision of, 25
urinary tract infection, treatment of, 64

Vacca, Joseph, 173–177
vaccines, x, xi, 8, 125–133, 194, 196
Vagelos, P. Roy, 160, 196
Valenzuela, Pablo, 196
Valium, 113, 122, 123
Vane, John, 153
Vasotec, 161, 175
venom, of pit viper, Brazilian, 153

vitamins, xi, 29–41, 76; B complex, 37; B$_1$, 33, 34; B$_2$ (riboflavin), 103, 122; B$_{12}$, 82, 117; C, 30, 32; deficiencies of, 29, 35, 37, 39; E, 35, 37; K, 38, 39; P, 32

Wadsworth, Augustus, 93
Waksman, Selman Abraham, 71, 78, 89, 90–92, 93, 141
Wallace Laboratories, 122
Walter Reed Army Institute of Research, 131
warfarin. *See* Coumadin
Waxman, Henry, 199
Webb, Lee, 150
Weigert, Carl, 17
Weissmann, Charles, 191–193

Welch, Arnold D., 151
Wellcome Physiological Research Laboratories, 59
Weller, Thomas, 126, 127, 133
Wellman, George, 150
Whitmore, Frank, 109
whooping cough vaccine, clinical trial of, 141
Wiley, Harvey Washington, 26–28
William S. Merrell Company, 142, 144, 164
Williams, Nina, 153
Williams, Robert R., 33–34
Windaus, Adolf, 35
Wolfenden, R. V., 153
Wong, David T., 167–168
Wong, Pancras, 154

Woodward-Hoffmann rules, 117
Woodward Research Institute, 117
Woodward, Robert Burns, 81, 114–117, 161
Wooyenaka, Keizo, 49
Wright, Almroth, 73
Wright, C. R. A., 12
Wyeth Laboratories, 127

X-ray crystallography, 82, 115

Zocor, 162
Zovirax, 172
Zyloprim, 172
Zyrtec, 164